SpringerBriefs in Mathematics

SpringerBriefs in Mathematics showcases expositions in all areas of mathematics and applied mathematics. Manuscripts presenting new results or a single new result in a classical field, new field, or an emerging topic, applications, or bridges between new results and already published works, are encouraged. The series is intended for mathematicians and applied mathematicians.

For further volumes:
http://www.springer.com/series/10030

Alberto Cabada

Green's Functions
in the Theory of Ordinary
Differential Equations

 Springer

Alberto Cabada
Department of Mathematical Analysis
University of Santiago de Compostela
Santiago de Compostela
Galicia, Spain

ISSN 2191-8198   ISSN 2191-8201 (electronic)
ISBN 978-1-4614-9505-5   ISBN 978-1-4614-9506-2 (eBook)
DOI 10.1007/978-1-4614-9506-2
Springer New York Heidelberg Dordrecht London

Library of Congress Control Number: 2013953341

Mathematics Subject Classification (2010): 34A30, 34B05, 34B08, 34B09, 34B10, 34B15, 34B27, 34C25, 34L05, 34L10, 34L15, 45A05, 45B05, 45C05, 45D05, 45Q05, 47A63, 47B07, 47B65

Printed on acid-free paper

Springer is part of Springer Science+Business Media (www.springer.com)

To my wife Marina and my sons Víctor and Martín

Preface

Ordinary differential equations involving additional conditions on the boundary of a given bounded interval have been exhaustively studied in the literature.

The classical results of Peano and Picard ensure, under suitable assumptions on the regularity of the nonlinear part of the equation, the existence and uniqueness of solutions of the considered initial value problem, which is defined in a neighborhood of the starting point.

It is important to point out that the aforementioned results are related to the concept of "local solution." However, if we are dealing with boundary value problems, the solution must be defined in the whole interval of definition and the local existence has no sense.

To see that the existence of solutions of boundary value problems cannot be deduced exclusively from the regularity data of the equation, it suffices to look for the 1-periodic solutions of the equation $u'(t) = 1$. It is obvious that the general solution of such equation is given by $u_c(t) = c + t$, with $c \in \mathbb{R}$, and none of them satisfies $u(0) = u(1)$. Thus, to ensure the existence of solutions of this kind of problems, we must take into account not only the regularity of the functions that appear in the equation but also the information provided by the boundary conditions.

Under suitable regularity assumptions on the linear operator L, we have that if L is a linear operator and equation $L\,u = f$, coupled with suitable homogeneous linear boundary value conditions on a real interval $[a, b]$, has only the trivial solution for $f \equiv 0$, then the associated linear operator is invertible and its inverse operator, $L^{-1} f$, is characterized by an integral kernel, $g(t, s)$, called Green's function,[1] and the solution of the considered problem is then given by

$$u(t) = L^{-1} f(t) := \int_a^b g(t, s)\, f(s)ds, \quad t \in [a, b].$$

---

[1] George Green (1793–1841) was the first mathematician to use such kind of kernels to solve boundary value problems.

We notice that, as it has been pointed out in [19], if we are able to obtain the expression of Green's kernel, we know the cases (if the linear operator depends on some parameters, for instance) in which it is not defined and, in consequence, the resonant cases (of nonuniqueness of the homogeneous problem) are explicitly given. The main advantage of Green's function is the fact that it is independent of the function f. To get the exact solution for each particular case of f we only need to calculate the corresponding integral, and so we have the expression that we are looking for (this is due to the fact that Green's function is just the kernel of the operator L^{-1}).

The aim of this monograph is to provide to graduate and doctoral students, together with researchers interested in this field, a comprehensive and thorough study of Green's functions. Along the monograph, some classical results of functional analysis are needed, which, for a better understanding of the text, will be introduced throughout the book as they are used. Along the chapters, some examples are given to illustrate the obtained results. Moreover, some particular cases are introduced to show the necessary conditions that are required to develop the theory.

The properties of these functions are widely used in the literature. But there are few books devoted just to the study of these functions [25, 27, 31, 38, 39, 53, 54]. In many cases, these functions are presented as the only function that verifies certain "a priori" given axioms. Our approach is posed differently; our intention is "to arrive" at Green's function. To do so, we consider, first, in Sect. 1.2, a linear system of first order, for which, using classical results of linear differential equations, we obtain the expression of the kernel of the integral equation that represents the solution we are looking for. This kernel is the so-called Green's function. Moreover, the determinant of a suitable matrix characterizes the uniqueness of such function. This determinant provides us the spectrum of the studied linear operator.

Although the paper is aimed to two-point boundary value problems, a characterization of how to use this theory when we are dealing with multipoint boundary conditions is also shown in Subsect. 1.2.1.

Once the characterization of the existence and uniqueness of Green's functions is presented, we prove some of their basic properties and their relationship to the related linear operator. For example, we will consider in Sect. 1.3 the relationship between the symmetric kernel of Green's function and the linear adjoint operator.

After this, we will focus on the scalar equations of nth order, for which, as in the previous case, we automatically get optimal existence and uniqueness conditions in Sect. 1.4. We also characterize the relationship between the existence and uniqueness of Green's function and the spectrum of the associated nth-order linear operator. Moreover, we will consider the study of the symmetry properties of Green's function and the self-adjoint character of the related linear operator. When the coefficients of the linear operator are constant, different techniques to calculate the exact expression of Green's functions are given. In the particular cases of initial, terminal, and periodic problems, the expression of Green's function follows from the inverse of a related constant matrix.

At this point, we introduce in Sect. 1.5 the method of lower and upper solutions. This tool is very well known in the theory of nonlinear boundary value problems. Only a few particular cases have been considered in this section. We introduce this method here to present different examples that point out the deep influence that the existence and uniqueness of Green's function of a related linear operator has on the existence of solution of nonlinear boundary value problems. More concisely, the existence results for nonlinear problems follow when there is a related Green's function with constant sign. So, it is fundamental to describe the cases in which the linear operators satisfy some suitable comparison principles, i.e., if the linear operator acting over a function has constant sign, then this function must have constant sign too.

These comparison principles are studied in Sect. 1.6. In this case the framework is very general and the concept of related set to a boundary condition is introduced. The equivalence of the validity of a comparison result in a particular related set and the constant sign of a Green's function is proved here. The validity of a comparison principle for a linear operator and for its adjoint is also pointed out.

Next section is devoted to monotone iterative techniques. As in the case of the lower and upper solutions, this is a tool used for nonlinear boundary value problems. In this monograph it is presented in a general framework for nth-order problems. This approach will be fundamental to Sect. 1.8, in which a one parameter family of nth-order linear operators is studied and the monotonicity dependence, with respect to the parameter, of the constant sign Green's functions is proved. Moreover, by using this kind of techniques, we describe the range of the parameters for which the linear operator has constant sign Green's function. This study is closely related to the spectral theory of completely continuous operators.

The last two sections are devoted to present the exact interval of the real parameter for which some particular Green's functions, related to given nth-order linear differential operators, have constant sign on the space of either the periodic or the separated boundary conditions.

The monograph ends with two appendices. The first one is concerned with the algorithm developed in [19] and implemented in a Mathematica program package. This program is of free access and can be downloaded from the web page of the author. It allows the effective calculation of Green's function provided that the differential operator has constant coefficients. These developments allow us to study their properties, both qualitative and quantitative, in a more accessible way. It is important to note that the main difficulty of this type of functions lies not only in its exact calculation but also even in the case where it can be performed, in the complexity of the expression obtained and, therefore, its handling becomes very difficult. Addressing methods of obtaining this expression more easily, we can make the study of its extreme values, its symmetry or some kind of boundedness, with less margin for error.

The second appendix shows the exact expression of a list of the most commonly studied operators. They have been calculated with the package explained previously and they are accessible from the Wolfram web page.

The monograph is completed with a bibliography of papers, both classic and recent, that have contributed to the development of this theory.

Santiago de Compostela, Spain   Alberto Cabada

Acknowledgements

I wish to express my sincere gratitude to my friends and colleagues who have encouraged me to make this monograph, specially to Beatriz Máquez-Villamarín for his help in processing this manuscript and José Ángel Cid, Rodrigo L. Pouso, and Adrián F. Tojo for their very careful reading of the manuscript, the discussions about many of the results that are recompiled here, and their interesting suggestions improving some of them.

Contents

Chapter 1
Green's Functions in the Theory of Ordinary Differential Equations

1.1 Preliminaries

In this monograph we will present the main topics concerning Green's functions related to nth-order ordinary differential equations coupled with linear two-point boundary conditions. To show the potential of this theory and importance of obtaining qualitative and quantitative properties of this kind of functions, we will consider in this preliminary section a simple example to illustrate the results we are dealing with.

It is very well known that, given $m \in \mathbb{R}$ and f a continuous function, if we consider the first-order scalar equation

$$u'(t) + m\,u(t) = f(t), \quad t \in \mathbb{R}, \tag{1.1.1}$$

it is enough to multiply both sides of the equation by factor $e^{m\,t}$ and to rewrite it in the equivalent form

$$\frac{d}{dt}\left(e^{m\,t}\,u(t)\right) = e^{m\,t}\,f(t), \quad t \in \mathbb{R}.$$

Now, by direct integration, we arrive to

$$e^{m\,t}\,u(t) - u(0) = \int_0^t e^{m\,s}\,f(s)\,ds, \quad t \in \mathbb{R},$$

or, which is the same, the set of solutions of (1.1.1) consists of the one-dimensional linear space:

$$u_c(t) = e^{-m\,t}\,c + \int_0^t e^{-m\,(t-s)}\,f(s)\,ds, \quad c \in \mathbb{R}. \tag{1.1.2}$$

A. Cabada, *Green's Functions in the Theory of Ordinary Differential Equations*, SpringerBriefs in Mathematics, DOI 10.1007/978-1-4614-9506-2_1,
© Alberto Cabada 2014

Of course, if we impose the additional initial value condition $u(0) = u_0$, we are in presence of a uniquely solvable problem, which unique solution is given by the expression (1.1.2) with $c = u_0$.

However, if we are interested in the solutions of (1.1.1) that are defined in the bounded interval $[0, 1]$ and satisfy that

$$u(0) = u(1), \tag{1.1.3}$$

we must note that in this case we do not know a priori the value of $u(0)$. So, the problem is equivalent to looking for, inside the one-dimensional set of solutions expressed in (1.1.2), what is the real parameter c for which the corresponding function u_c attains the same value at the beginning and at the end of the interval of definition. In that case, we may ask not only if such a value exists, but also must verify if it is, or not, unique and, which is the data involved in the equation (parameter m, nonhomogeneous function f, interval of definition $[0, 1]$) that characterize these properties.

In our case, from (1.1.2) we have that (1.1.3) holds if and only if

$$c_m = \frac{1}{1 - e^{-m}} \int_0^1 e^{-m(1-s)} f(s) \, ds, \tag{1.1.4}$$

that is, if $m \neq 0$, we have that there exists a unique solution of problem (1.1.1), (1.1.3) and the attained value of such solution at the extremes of the interval is given by the expression (1.1.4).

Of course, the expression of the unique solution u depends on the values of the nonhomogeneous part f, but it is clear that the uniqueness of solution property only depends on the values of the real parameter m and that function f has no influence on this fact. We only need to assume on f regularity enough to ensure that the integrals in (1.1.2) are well defined.

When $m = 0$, expression (1.1.4) has no sense. In such a case, by direct integration, we deduce that a necessary condition to ensure the existence of solution is that $\int_0^1 f(s) \, ds = 0$. It is obvious that under this assumption on the nonhomogeneous function f, there are infinitely many solutions of problem (1.1.1), (1.1.3). In fact the set of these solutions is given by the one-dimensional linear space

$$u_c(t) = c + \int_0^t f(s) \, ds, \quad c \in \mathbb{R}.$$

This example shows us that if we are in presence of a nonhomogeneous linear boundary value problem, the situation concerning the corresponding set of solutions is similar to a nonhomogeneous algebraic matrix equation $A x = b$. If the square matrix A is invertible, we have a unique solution $x = A^{-1} b$. The value of this unique solution x depends on b, but the uniqueness character is only related to the invertibility of the matrix A and has no dependence on the expression of the vector b. However, when there is no inverse of A, the existence of solutions depends on the nonhomogeneous term b. In such a case, there is a solution of the matrix equation if and only if the rank of A and the one of its enlarged matrix $(A \,|\, b)$ coincide.

Moreover, if this is the case, there are infinitely many solutions that form a finite dimensional affine space.

In our situation we can think of our problem (1.1.1), (1.1.3) as follows.

Let $X = \{u \in \mathscr{C}^1([0, 1], \mathbb{R}), u(0) = u(1)\}$ be the Banach space equipped with the norm $\|u\| = \max\{\|u\|_\infty, \|u'\|_\infty\}$ and let $Y = \mathscr{C}([0, 1], \mathbb{R})$ equipped with the supremum norm. Consider, for any $m \in \mathbb{R}$, the linear operator $L_m : X \to Y$ defined as $L_m u := u' + m u$.

Given $f \in Y$, find those functions $u_m \in X$, such that $L_m u_m = f$.

If we have a unique solution u_m, we can write it as $u_m = L_m^{-1} f$. Where $L_m^{-1} : Y \to X$ is defined as the operator such that to any function $f \in Y$ assigns the unique solution $u_m \in X$ of equation $L_m u_m = f$. In other words, the uniqueness of solution is equivalent to the study of the values of the real parameter m for which $L_m : X \to Y$ is invertible.

Concerning the nature of the inverse operator of L_m in X we have that as opposed to matrix calculus, in which the inverse of a matrix is a matrix too, if we have a linear differential operator, its inverse is given by an integral one. This is not very surprising, since in order to solve an equation that involves the derivatives of a given function, we must use integration theory.

Returning to our example, by pasting in (1.1.2) the value of the parameter c_m given in (1.1.4), we have that the unique solution of problem (1.1.1), (1.1.3) (with $m \neq 0$) is given by the following expression:

$$u_m(t) = \frac{e^{-mt}}{1 - e^{-m}} \int_0^1 e^{-m(1-s)} f(s)\,ds + \int_0^t e^{-m(t-s)} f(s)\,ds$$

$$= \frac{e^{-mt}}{1 - e^{-m}} \int_0^1 e^{-m(1-s)} f(s)\,ds + \int_0^1 e^{-m(t-s)} \chi_{(0,t)}(s)\, f(s)\,ds$$

$$= \int_0^1 g_m(t, s)\, f(s)\,ds =: L_m^{-1} f(t),$$

where $\chi_{(0,t)}$ is the indicator function of the interval $(0, t)$ and

$$g_m(t, s) = \frac{1}{1 - e^{-m}} \begin{cases} e^{-m(t-s)}, & \text{if } 0 < s < t < 1, \\ e^{-m(1+t-s)}, & \text{if } 0 < t < s < 1. \end{cases}$$

The obtained function G_m is the so-called *Green's function* related to the boundary value problem (1.1.1), (1.1.3). Such function is the kernel of the inverse operator L_m^{-1} and gives us crucial information about the solutions of the equation. The first one is that in a high number of situations we can obtain the exact expression of the solutions by direct integration. For instance, if $f(t) = t$, we have that

$$u_m(t) = \frac{e^{m(1-t)}}{m(e^m - 1)} + \frac{t}{m} - \frac{1}{m^2}.$$

When $f(t) = \cos^2(\pi t)$, the unique solution is given by

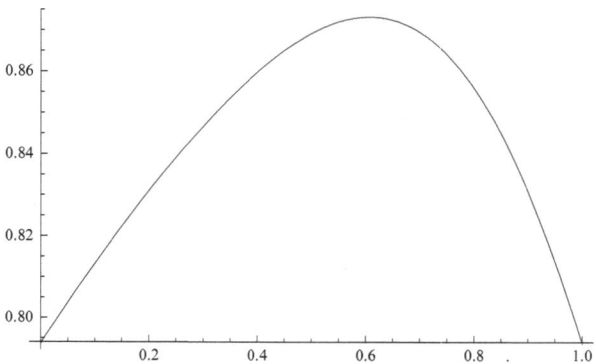

Fig. 1.1 Approximate solution of the problem $u'(t) + u(t) = e^{-t^4}$, $u(0) = u(1)$

$$u_m(t) = \frac{m^2 \cos(2\pi t) + m^2 + 2\pi m \sin(2\pi t) + 4\pi^2}{2m^3 + 8\pi^2 m}.$$

When the calculus of antiderivatives fails and we are not able to calculate the exact expression of the integral for a given function f, we can use numerical integration and so plot the graph of the desired function. This is the case, for instance, when we consider, in problem (1.1.1), (1.1.3), the parameter $m = 1$ and the function $f(t) = e^{-t^4}$. We cannot solve the integral that gives us the exact expression of the solution, but we are able to solve it numerically and plot the approximate graph as we do in Fig. 1.1.

However, the most important information that Green's function of a related boundary value problem gives us is qualitative. In our example it is immediate to verify that $m\, g_m(t, s) > 0$ for all $t, s \in [0, 1]$ and $m \neq 0$. As a consequence, we deduce the following comparison principle:

Let $u \in X$ be such that $L_m u \geq 0$ in $[0, 1]$. Then $m\, u \geq 0$ in $[0, 1]$.

Notice that, as a direct consequence, if $f_1 \geq f_2$ in $[0, 1]$, then the corresponding solutions satisfy $m\, u_1 \geq m\, u_2$ in $[0, 1]$.

Moreover, one can verify that, for any $t, s \in [0, 1]$ fixed, Green's function g_m is strictly decreasing (and positive) with respect to $m \in (0, \infty)$. So, if we consider $0 < m_1 < m_2$ and $f_1 \geq f_2 \geq 0$ in $[0, 1]$, we deduce that for all $t \in [0, 1]$ the corresponding solutions in X of the equation $L_{m_i} u_i = f_i, i = 1, 2$, are nonnegative on $[0, 1]$ and satisfy

$$u_1(t) = \int_0^1 g_{m_1}(t, s)\, f_1(s)\, ds \geq \int_0^1 g_{m_2}(t, s)\, f_1(s)\, ds \geq \int_0^1 g_{m_2}(t, s)\, f_2(s)\, ds = u_2(t).$$

For negative values of m we have that Green's function g_m is strictly decreasing (and negative) with respect to m. So, in an analogous way, we deduce that if $m_1 < m_2 < 0$ and $f_2 \geq f_1 \geq 0$ in $[0, 1]$, then $u_2(t) \leq u_1(t) \leq 0$ for all $t \in [0, 1]$.

On the other hand, it is immediate to verify that

$$\int_0^1 g_m(t,s)\, ds = \frac{1}{m}$$

and

$$g_m(t,t^+) = \frac{1}{e^m - 1} \le g_m(t,s) \le \frac{e^m}{e^m - 1} = g_m(t,t^-), \quad \text{for all } t,s \in [0,1].$$

Therefore

$$\text{ess sup}_{t,s \in [0,1]}\{|g_m(t,s)|\} = \frac{\max\{1,\, e^m\}}{|e^m - 1|}.$$

As a consequence, we deduce the following a priori estimates for the solution $u_m \in X$ of the equation $L_m u = f$:

$$\|u_m\|_\infty \le \|f\|_\infty / |m| \quad \text{and} \quad \|u_m\|_\infty \le \frac{\max\{1,\, e^m\}}{|e^m - 1|}\|f\|_1.$$

By means of the Hölder's inequality we can generalize this last estimate to the following one (denoting, as usual, $(p/(p-1) = \infty$ if $p = 1$; $p/(p-1) = 1$ if $p = \infty$):

$$\|u_m\|_\infty \le \frac{\left|\frac{e^{mp}-1}{mp}\right|^{\frac{1}{p}}}{|e^m - 1|}\|f\|_{p/(p-1)}, \quad p \ge 1.$$

Finally, it is important to notice that we can work with nonhomogeneous boundary conditions. For instance, we can look for a solution of (1.1.1) that satisfies the nonhomogeneous periodic conditions $u(0) - u(1) = \lambda \in \mathbb{R}$. In this case we must solve the homogeneous case (1.1.1), (1.1.3), for which we know the solution by means of Green's function, and add its solution to one of the problem:

$$v'(t) + m\,v(t) = 0, \ t \in [0,1], \quad v(0) - v(1) = \lambda.$$

By trivial computations we know that

$$v(t) = \lambda \frac{e^{-mt}}{1 - e^{-m}} = \lambda\, g_m(t,0).$$

So, we can extend the comparison principle given in the space X to the more general one $Z = \{u \in \mathscr{C}^1([0,1],\mathbb{R}),\ u(0) \ge u(1)\}$. Obviously, the a priori bounds can be automatically adapted to this new situation.

This simple example is just an overview of the importance of the quantitative and qualitative properties that we can deduce from Green's function of a considered equation. There are many more related to the existence of solutions of nonlinear problems. Some of them will be presented along this monograph.

Although the example given in this preliminary section is coupled with a boundary condition that involves the values of the function at the extremes of the interval (this is known as a two-point boundary condition) these additional conditions could be given instead in some interior points of the interval ($u(1/2) = 0$) or involve different points that belong to the interior and the boundary of the interval ($u(0) + u(1/3) + u(1) = 0$). In this last case the boundary conditions are known as multipoint boundary conditions.

Other kinds of linear boundary conditions may consider all the values attained by the function in the whole interval (or in a part of it): $\int_a^b u(s)\,ds = u(\eta)$, with $[a,b] \subset [0,1]$ and $\eta \in [0,1]$. These kinds of boundary conditions are known as nonlocal.

In this work we will pay special attention to the two-point boundary conditions, but the general theory for this kind of problems can be adapted to the nonlocal and multipoint cases.

1.2   Construction of Green's Function

Our work will be dedicated to the study of nth-order ordinary differential equations with additional conditions concerning the extremes of the interval. Since such equations are equivalent to a suitable n-dimensional first-order differential linear system, we will present the general settings of this theory inside this more general framework.

Thus, we are interested in solving the following first-order differential n-dimensional linear system

$$x'(t) = A(t)\,x(t) + f(t), \quad t \in J = [a,b], \tag{1.2.1}$$

together with the two-point boundary value conditions

$$B\,x(a) + C\,x(b) = h. \tag{1.2.2}$$

Here, n is a positive integer, $a,\ b \in \mathbb{R}$, $a < b$, $A : J \longrightarrow \mathcal{M}_{n \times n}$ is a $\mathcal{L}^1(J, \mathcal{M}_{n \times n})$ function, $f : J \longrightarrow \mathbb{R}^n$ belongs to $\mathcal{L}^1(J, \mathbb{R}^n)$, $B, C \in \mathcal{M}_{n \times n}$ and $h \in \mathbb{R}^n$ have constants coefficients, and $x : J \longrightarrow \mathbb{R}^n$ belongs to the set $\mathscr{AC}(J, \mathbb{R}^n)$. As usual, we denote by $\mathcal{L}^1(J, \mathbb{R}^n)$ and $\mathcal{L}^1(J, \mathcal{M}_{n \times n})$ the set of all Lebesgue integrable functions on J and by $\mathscr{AC}(J, \mathbb{R}^n)$ the set of absolutely continuous functions on J.

Clearly, n, a, b, A, f, B, C, and h are the known data of the problem and x is the unknown variable.

First, we study the structure of the set of solutions of the homogeneous problem ($f \equiv 0$, $h \equiv 0$):

$$x'(t) = A(t) x(t), \quad t \in J, \quad B x(a) + C x(b) = 0. \qquad (1.2.3)$$

Let $W = \{x \in \mathscr{AC}(J, \mathbb{R}^n); \ B x(a) + C x(b) = 0\}$ and define the linear operator

$$L : x \in W \longrightarrow L x = x' - A x \in \mathscr{L}^1(J, \mathbb{R}^n). \qquad (1.2.4)$$

As a consequence, the set of solutions of system (1.2.3) coincides with the kernel of operator L. So, we have that the set of solutions of problem (1.2.3) is a linear space of dimension $k \le n$.

The "natural" question that we are dealing with is: What is the exact value of such dimension? In other words: What is the influence of the data of the considered problem (A, B, C, and the interval J) on the dimension of the set of solutions of the homogeneous problem (1.2.3)?

Before answering this question, we present an example that could illustrate this problem.

Example 1.2.1. For any $\lambda \in \mathbb{R}$, consider the following first-order two-dimensional problem:

$$\begin{pmatrix} x'(t) \\ y'(t) \end{pmatrix} = \begin{pmatrix} 0 & 1 \\ -\lambda & 0 \end{pmatrix} \begin{pmatrix} x(t) \\ y(t) \end{pmatrix}, \quad t \in [0, 2\pi],$$

$$\begin{pmatrix} 0 & 1 \\ 0 & 0 \end{pmatrix} \begin{pmatrix} x(0) \\ y(0) \end{pmatrix} + \begin{pmatrix} 0 & 0 \\ 0 & 1 \end{pmatrix} \begin{pmatrix} x(2\pi) \\ y(2\pi) \end{pmatrix} = \begin{pmatrix} 0 \\ 0 \end{pmatrix}.$$

Clearly, the x part of the solution solves the second-order equation

$$x''(t) + \lambda x(t) = 0, \quad t \in [0, 2\pi], \quad x'(0) = x'(2\pi) = 0, \qquad (1.2.5)$$

and $y(t) = x'(t)$.

When $\lambda < 0$, we have that function x is given by the general expression

$$x(t) = C_1 e^{\sqrt{-\lambda} t} + C_2 e^{-\sqrt{-\lambda} t}, \quad \text{for } C_1, C_2 \in \mathbb{R}.$$

Such expression fulfills the boundary conditions if and only if

$$\begin{pmatrix} C_1 \\ C_2 \end{pmatrix} \in \ker \left\{ \begin{pmatrix} 1 & -1 \\ e^{2\pi\sqrt{-\lambda}} & -e^{-2\pi\sqrt{-\lambda}} \end{pmatrix} \right\} = \left\{ \begin{pmatrix} 0 \\ 0 \end{pmatrix} \right\}.$$

As a consequence, if $\lambda < 0$, we deduce that the considered problem has only the trivial solution. That is, $\ker(L) = \left\{ \begin{pmatrix} 0 \\ 0 \end{pmatrix} \right\}$, where L is the operator

$$L \begin{pmatrix} x(\cdot) \\ y(\cdot) \end{pmatrix} = \begin{pmatrix} x'(\cdot) \\ y'(\cdot) \end{pmatrix} - \begin{pmatrix} 0 & 1 \\ -\lambda & 0 \end{pmatrix} \begin{pmatrix} x(\cdot) \\ y(\cdot) \end{pmatrix}$$

defined on the space

$$W = \left\{ \begin{pmatrix} x \\ y \end{pmatrix} \in \mathscr{AC}\left([0, 2\pi], \mathbb{R}^2\right) : \begin{pmatrix} 0 & 1 \\ 0 & 0 \end{pmatrix} \begin{pmatrix} x(0) \\ y(0) \end{pmatrix} + \begin{pmatrix} 0 & 0 \\ 0 & 1 \end{pmatrix} \begin{pmatrix} x(2\pi) \\ y(2\pi) \end{pmatrix} = \begin{pmatrix} 0 \\ 0 \end{pmatrix} \right\}.$$

When $\lambda = 0$, we know that

$$x(t) = C_1 + C_2 t.$$

The boundary condition is fulfilled if and only if

$$C_2 = 0.$$

Therefore, when $\lambda = 0$, every constant solution x solves (1.2.5).
So $\dim(\ker L) = 1$ and the solutions of the system consist on the one-dimensional linear space generated by the vector function $\begin{pmatrix} 1 \\ 0 \end{pmatrix}$.

When $\lambda > 0$, the expression of the function x is given by

$$x(t) = C_1 \cos(\sqrt{\lambda}\, t) + C_2 \sin(\sqrt{\lambda}\, t).$$

The boundary condition is fulfilled if and only if $C_2 = 0$ and $C_1 \sin(2\sqrt{\lambda}\,\pi) = 0$.
If $\lambda \neq \dfrac{n^2}{4}$ for all $n \in \mathbb{N}$, we have that the unique solution of (1.2.5) is the zero solution and, as a consequence, operator L is injective.

When $\lambda = \dfrac{n^2}{4}$ for some $n \in \{1, 2, \ldots\}$ we have nontrivial solutions of the form

$$x(t) = C_1 \cos\frac{n}{2} t, \qquad C_1 \in \mathbb{R}.$$

Thus, $\ker L$ is the one-dimensional linear space generated by $\begin{pmatrix} \cos\frac{n}{2} t \\ -\frac{n}{2} \sin\frac{n}{2} t \end{pmatrix}$.

Notice that we have a sequence of values of λ ($\lambda_n = \dfrac{n^2}{4}$, $n = 0, 1, \ldots$) for which the problem has nontrivial solutions. So, we have a real number λ_n (eigenvalue) and a related linear space of dimension one generated by $\begin{pmatrix} \cos\frac{n}{2} t \\ -\frac{n}{2} \sin\frac{n}{2} t \end{pmatrix}$ (eigenfunction) consisting of the nontrivial solutions of the considered problem.

The previous example shows that on the contrary to the initial value problem, in general we cannot expect problem (1.2.3) to have only the trivial solution.

We remark that problem (1.2.3) covers the initial value problem when B is the identity matrix and $C \equiv 0$. The regularity of the matrix function A guarantees that the unique solution of the initial value problem is defined in the whole interval J [28, p. 47].

Next, we give a link between the solutions of the homogeneous problem (1.2.3) and the general one (1.2.1)–(1.2.2).

Theorem 1.2.2. *x is a solution of problem (1.2.1)–(1.2.2) if and only if $x = y + p$, where y is a solution of the homogeneous problem (1.2.3) and p is a solution of (1.2.1)–(1.2.2).*

Proof. Let y be a solution of (1.2.3) and p be a solution of problem (1.2.1)–(1.2.2). As a consequence

$$y'(t) + p'(t) = A(t)\, y(t) + A(t)\, p(t) + f(t) = A(t)\, (y(t) + p(t)) + f(t)$$

and $x \equiv y + p$ fulfills system (1.2.1) on J.

On the other hand,

$$B(y + p)(a) + C(y + p)(b) = B\, p(a) + C\, p(b) = h,$$

and x is a solution of (1.2.1)–(1.2.2).

Consider now x_1 and x_2, two solutions of problem (1.2.1)–(1.2.2). As a consequence

$$x_1'(t) - x_2'(t) = A(t)\, (x_1(t) - x_2(t)) \qquad \text{for all } t \in J$$

and

$$B(x_1(a) - x_2(a)) + C(x_2(b) - x_1(b)) = h - h = 0.$$

That is, the difference of two solutions of problem (1.2.1)–(1.2.2) is a solution of the homogeneous problem (1.2.3). Thus the proof is concluded. $\qquad\square$

Let $\phi : J \longrightarrow \mathcal{M}_{n \times n}$ be a fundamental matrix related to system (1.2.3), i.e., ϕ is regular functional matrix and a solution of the linear matrix equation:

$$\phi'(t) = A(t)\, \phi(t), \qquad t \in J. \tag{1.2.6}$$

We arrive at the following existence and uniqueness result for problem (1.2.1)–(1.2.2).

Theorem 1.2.3. *Problem (1.2.1)–(1.2.2) has a unique solution $x \in \mathscr{AC}(J, \mathbb{R}^n)$ if and only if*

$$\det\left(M_\phi\right) \neq 0, \tag{1.2.7}$$

with ϕ any fundamental matrix of system (1.2.3) and $M_\phi \equiv B\, \phi(a) + C\, \phi(b)$.

Proof. From the variation of constants formula [28, Corollary 2.1], we have that $x \in \mathscr{AC}(J, \mathbb{R}^n)$ is a solution of problem (1.2.1) if and only if there exists $\lambda \in \mathbb{R}^n$ such that

$$x(t) = \phi(t)\lambda + \phi(t) \int_a^t \phi^{-1}(s) f(s) \, ds, \qquad t \in J. \qquad (1.2.8)$$

Obviously, function x satisfies the boundary value condition (1.2.2) if and only if λ solves the following algebraic equation:

$$M_\phi \lambda \equiv (B \phi(a) + C \phi(b)) \lambda = h - C \phi(b) \int_a^b \phi^{-1}(s) f(s) \, ds. \qquad (1.2.9)$$

It is clear that this equation has a unique solution if and only if matrix M_ϕ is invertible. □

As a consequence, we deduce the following result.

Corollary 1.2.4. *Problem (1.2.3) has only the trivial solution if and only if problem (1.2.1)–(1.2.2) has a unique solution.*

Remark 1.2.5. Notice that both the initial and the terminal value problems are uniquely solvable. To see this it is enough to take into account that the initial problem corresponds to the choice $B = I_n$ and $C = 0$ ($M_\phi = \phi(a)$) and the terminal one is given by $B = 0$ and $C = I_n$ ($M_\phi = \phi(b)$). Here we denote by I_n the n-dimensional identity matrix.

Remark 1.2.6. Due to the fact that given two fundamental matrices ϕ and ψ of problem (1.2.3) there is a nonsingular $n \times n$ constant matrix D such that $\phi(t) = \psi(t) D$ [20, Theorem 2.3], we have that

$$M_\phi = B \phi(a) + C \phi(b) = (B \psi(a) + C \psi(b)) D = M_\psi D$$

and, as a consequence, condition (1.2.7) is independent of the chosen fundamental matrix.

Remark 1.2.7. Notice that when condition (1.2.7) holds, the expression of the unique solution of problem (1.2.1)–(1.2.2) is given by

$$x(t) = \phi(t) M_\phi^{-1} \left(h - C \phi(b) \int_a^b \phi^{-1}(s) f(s) \, ds \right) + \phi(t) \int_a^t \phi^{-1}(s) f(s) \, ds.$$

$$(1.2.10)$$

Obviously, such expression depends on the different data of the system. However, the uniqueness of solutions is ensured without influence of the nonhomogeneous parts f and h.

On condition (1.2.7), only the matrices B and C, the extremes of the interval J and the fundamental matrix, which depends on the matrix function $A(t)$, appear. As in Remark 1.2.6 it is immediate to verify that the expression in (1.2.10) is independent of the choice of the fundamental matrix ϕ.

Remark 1.2.8. In case (1.2.7) does not hold, we know that problem (1.2.1)–(1.2.2) is solvable if and only if the algebraic system (1.2.9) is solvable too.

Such property is possible only in the case of rank (M_ϕ) coincides with the rank of the matrix $\left(M_\phi \,\middle|\, h - C\,\phi(b) \int_a^b \phi^{-1}(s)\, f(s)\, ds \right)$.

In this situation, on the contrary to the uniqueness of solutions case, the existence (or not) of solutions depends on the nonhomogeneous parts of the system: f and h.

Clearly, if the set of solutions is not empty, it consists of an affine vector space of dimension $n - \mathrm{rank}\,(M_\phi)$. From Remark 1.2.6, this rank is independent from the choice of the fundamental matrix ϕ.

Remark 1.2.9. It is important to remark that to ensure the uniqueness of solution of problem (1.2.1)–(1.2.2) for any f in $\mathscr{L}^1(J, \mathbb{R}^n)$ and $h \in \mathbb{R}^n$, the involved boundary conditions (1.2.2) must define n-linearly independent conditions. So we obtain the following necessary condition:

$$\mathrm{rank}\,(B \mid C) = n. \tag{1.2.11}$$

However, due to the fact that this condition does not impose any restriction in the matrix function $A(t)$, it is not sufficient. To see this, it suffices to think about the homogeneous problem (1.2.3), with $A(t) = 0$, coupled with the periodic boundary conditions $B = -C = I_n$. It is immediate to verify that in this case condition (1.2.11) is fulfilled. However, any constant vector in $\mathbb{R}^n$ is a solution of (1.2.3).

Having in mind the previous remark, we are interested in obtaining a characterization of the uniqueness of solutions for problem (1.2.1)–(1.2.2) that involves condition (1.2.11). To this end, we must take into account that the general solution of the differential system (1.2.1) is given by (1.2.8) or, alternatively, by

$$x(t) = \phi(t)\lambda + \phi(t) \int_{t_0}^t \phi^{-1}(s) f(s)\, ds, \quad \lambda \in \mathbb{R}^n, \tag{1.2.12}$$

where $t_0 \in J$ can be chosen as we please.

For later purposes, it will be convenient to fix $t_0 \in (a, b)$, and then the solution x given by (1.2.12) is a solution of (1.2.1)–(1.2.2) if and only if $\lambda \in \mathbb{R}^n$ solves the algebraic system

$$M_\phi \lambda = h - B\phi(a) \int_{t_0}^a \phi^{-1}(s) f(s)\, ds - C\phi(b) \int_{t_0}^b \phi^{-1}(s) f(s)\, ds, \tag{1.2.13}$$

where M_ϕ is given in Theorem 1.2.3.

Next we present the following characterization of the uniqueness of solutions of problem (1.2.1)–(1.2.2) by means of the condition (1.2.11).

Theorem 1.2.10. *Let* $L : W \longrightarrow \mathscr{L}^1(J, \mathbb{R}^n)$ *be the linear operator defined in* (1.2.4) *and let* $\phi : J \longrightarrow \mathscr{M}_{n \times n}$ *be a fundamental matrix of the homogeneous system* (1.2.3).

The following statements are equivalent:

(i) *Problem (1.2.1)–(1.2.2) has a unique solution $x \in \mathscr{AC}(J, \mathbb{R}^n)$.*
(ii) $\det(M_\phi) \neq 0$.
(iii) *L is bijective.*
(iv) *L is injective.*
(v) *L is surjective and condition (1.2.11) holds.*

Proof. Statements (i), (ii), and (iii) are equivalent by virtue of Theorem 1.2.3 and the fact that (iii) implies (iv) is trivial.

Let us see that (iv) implies (ii). First, notice that (iv) ensures that $u = 0$ is the unique solution of (1.2.3), and this, in turn, is equivalent to saying that $\lambda = 0 \in \mathbb{R}^n$ is the unique solution of the algebraic system $M_\phi \lambda = 0$, which is equivalent to (ii).

So we have that the four first assertions are equivalent.

Now, we show that they are equivalent to property (v). To see this, assume condition (iii), in particular, L is surjective and the first part of (v) is proven.

Let us prove now that (iii) implies (1.2.11). We have already shown that (iii) implies (ii), so for every $w \in \mathbb{R}^n$ there exists $\lambda \in \mathbb{R}^n$ such that $M_\phi \lambda = w$, i.e.,

$$B\phi(a)\lambda + C\phi(b)\lambda = w.$$

This implies that the linear mapping

$$(v_1, v_2) \in \mathbb{R}^n \times \mathbb{R}^n \longmapsto (B|C) \begin{pmatrix} v_1 \\ v_2 \end{pmatrix} = Bv_1 + Cv_2 \in \mathbb{R}^n,$$

is onto, which is equivalent to (1.2.11)

Finally, we prove that (v) implies (ii). Condition (1.2.11) guarantees that for a fixed $w \in \mathbb{R}^n$ there exist $v_1, v_2 \in \mathbb{R}^n$ such that $Bv_1 + Cv_2 = -w$. We fix $t_0 \in (a, b)$ and we define an integrable function by

$$f(t) = \begin{cases} \dfrac{1}{a - t_0}\phi(t)\phi^{-1}(a)v_1, & \text{if } a \leq t \leq t_0, \\[3mm] \dfrac{1}{b - t_0}\phi(t)\phi^{-1}(b)v_2, & \text{if } t_0 < t \leq b. \end{cases}$$

Assumption (v) guarantees that the corresponding problem (1.2.1)–(1.2.2) with $h = 0$ has at least one solution x or, equivalently, there is at least one $\lambda \in \mathbb{R}^n$ which fulfills (1.2.13) with $h = 0$. Substituting the expression of f into (1.2.13) with $h = 0$ gives

$$M_\phi \lambda = -Bv_1 - Cv_2 = w.$$

Since w was fixed arbitrarily in $\mathbb{R}^n$, we deduce that the matrix M_ϕ has maximum rank, which implies (ii). $\qquad\square$

As we will see in Example 1.2.19, operator L can be surjective and not injective when (1.2.11) is not satisfied. However, L is necessarily surjective when it is injective.

In the next examples we apply the previous results to study the existence and uniqueness of some boundary value problems.

Example 1.2.11. Given $h_1, h_2 \in \mathbb{R}$, obtain the expression of the solutions of the following system:

$$\begin{cases} x'(t) = y(t), & t \in [0, 2\pi], \ x(0) = h_1, \\ y'(t) = -x(t), \ t \in [0, 2\pi], \ y(2\pi) = h_2. \end{cases}$$

The system fulfills the form (1.2.1)–(1.2.2) for

$$A(t) = \begin{pmatrix} 0 & 1 \\ -1 & 0 \end{pmatrix}, \ f(t) = \begin{pmatrix} 0 \\ 0 \end{pmatrix}, \ B = \begin{pmatrix} 1 & 0 \\ 0 & 0 \end{pmatrix}, \ C = \begin{pmatrix} 0 & 0 \\ 0 & 1 \end{pmatrix}, \ h = \begin{pmatrix} h_1 \\ h_2 \end{pmatrix}.$$

Since A is a constant function, we know that e^{At} is a fundamental matrix of the system, that is,

$$\phi(t) = \begin{pmatrix} \cos t & \sin t \\ -\sin t & \cos t \end{pmatrix}.$$

Since

$$M_\phi = B\,\phi(0) + C\,\phi(2\pi) = \begin{pmatrix} 1 & 0 \\ 0 & 1 \end{pmatrix},$$

we have that condition (1.2.7) is fulfilled and, as a consequence, this problem has a unique solution for every $h_1, h_2 \in \mathbb{R}$.
   Moreover, (1.2.10) tells us that

$$x(t) = h_1 \cos t + h_2 \sin t$$

and

$$y(t) = -h_1 \sin t + h_2 \cos t.$$

Example 1.2.12. Obtain the expression of the solutions of the following problem:

$$u''(t) + u(t) = 0, \ t \in [0, 2\pi], \quad u(0) + u(2\pi) = 1.$$

Denoting $x(t) = u(t)$ and $y(t) = u'(t)$, we arrive to an equivalent system defined by

$$A(t) = \begin{pmatrix} 0 & 1 \\ -1 & 0 \end{pmatrix}, \ f(t) = \begin{pmatrix} 0 \\ 0 \end{pmatrix}, \ B = \begin{pmatrix} 1 & 0 \\ 0 & 0 \end{pmatrix}, \ C = \begin{pmatrix} 1 & 0 \\ 0 & 0 \end{pmatrix}, \ h = \begin{pmatrix} 1 \\ 0 \end{pmatrix}.$$

In this case rank $(B \,|\, C) = 1$; thus Theorem 1.2.10 ensures that operator L is not bijective on W. Moreover

$$\phi(t) = \begin{pmatrix} \cos t & \sin t \\ -\sin t & \cos t \end{pmatrix} \text{ and } M_\phi = B\,\phi(0) + C\,\phi(2\,\pi) = \begin{pmatrix} 2 & 0 \\ 0 & 0 \end{pmatrix}.$$

In consequence, condition (1.2.7) does not hold (as we previously know, from Theorem 1.2.10). So, this problem is solvable if and only if system (1.2.9), rewritten in this case as

$$\begin{pmatrix} 2 & 0 \\ 0 & 0 \end{pmatrix} \begin{pmatrix} \lambda_1 \\ \lambda_2 \end{pmatrix} = \begin{pmatrix} 1 \\ 0 \end{pmatrix},$$

admits solutions too. In this case the solutions are given by $\begin{pmatrix} 1/2 \\ \lambda_2 \end{pmatrix}$, $\lambda_2 \in \mathbb{R}$.

Consequently, we deduce from (1.2.8) that

$$u(t) = \frac{1}{2}\,\cos t + \lambda_2\,\sin t, \ \lambda_2 \in \mathbb{R},$$

which forms a one-dimensional (two minus the rank of M_ϕ) affine vector space.

Example 1.2.13. Given $h_1, h_2 \in \mathbb{R}$, obtain the solutions of the problem:

$$u''(t) + u(t) = 0, \ t \in [0, 2\,\pi], \quad u(0) = u(2\,\pi) + h_1, \ u'(0) = u'(2\,\pi) + h_2.$$

In this situation, arguing as in the previous examples, we have that this problem is equivalent to the system (1.2.1)–(1.2.2), with

$$A(t) = \begin{pmatrix} 0 & 1 \\ -1 & 0 \end{pmatrix}, \ f(t) = \begin{pmatrix} 0 \\ 0 \end{pmatrix}, \ B = \begin{pmatrix} 1 & 0 \\ 0 & 1 \end{pmatrix}, \ C = \begin{pmatrix} -1 & 0 \\ 0 & -1 \end{pmatrix}, \ h = \begin{pmatrix} h_1 \\ h_2 \end{pmatrix}.$$

Moreover, $\phi(t) = \begin{pmatrix} \cos t & \sin t \\ -\sin t & \cos t \end{pmatrix}$ and $M_\phi = \begin{pmatrix} 0 & 0 \\ 0 & 0 \end{pmatrix}$.

Therefore, this problem has at least one solution if and only if $h_1 = h_2 = 0$. In such a case, (1.2.9) has solution $\begin{pmatrix} \lambda_1 \\ \lambda_2 \end{pmatrix}$, for all $\lambda_1, \lambda_2 \in \mathbb{R}$, and the solutions that we are looking for are given by the expression

$$u(t) = \lambda_1\,\cos t + \lambda_2\,\sin t.$$

In this case we have a two-dimensional (two minus the rank of M_ϕ) linear space of solutions generated by $\cos t$ and $\sin t$.

By virtue of Theorem 1.2.3, we know that condition (1.2.7) is a necessary and sufficient condition to ensure that problem (1.2.1)–(1.2.2) has a unique solution. In such a case, the unique solution $x \in \mathscr{AC}(J, \mathbb{R}^n)$ is given by expression (1.2.10).

Notice that considering $\chi_{(0,t)}$ the indicator function in $(0, t)$, equality (1.2.10) can be rewritten as follows:

$$x(t) = \phi(t)\, M_\phi^{-1} \left(h - C\, \phi(b) \int_a^b \phi^{-1}(s)\, f(s)\, ds \right) + \phi(t) \int_a^b \phi^{-1}(s)\, \chi_{(0,t)}(s)\, f(s)\, ds,$$

or, which is the same,

$$x(t) = \int_a^b G(t, s)\, f(s)\, ds + \phi(t)\, M_\phi^{-1}\, h, \qquad (1.2.14)$$

with

$$G(t, s) = \begin{cases} -\phi(t)\, M_\phi^{-1}\, C\, \phi(b)\, \phi^{-1}(s) + \phi(t)\, \phi^{-1}(s), & \text{if } a \le s < t \le b, \\[2mm] -\phi(t)\, M_\phi^{-1}\, C\, \phi(b)\, \phi^{-1}(s), & \text{if } a \le t < s \le b. \end{cases}$$

$$(1.2.15)$$

The function $G : (J \times J) \backslash \{(t, t),\ t \in J\} \longrightarrow \mathcal{M}_{n \times n}$ is called the **Green's function related to problem (1.2.3)**.

As we see in expression (1.2.15), Green's function $G \equiv (G_{i,j})$, $i,\ j \in \{1, \dots, n\}$ is not defined in the diagonal of the square $J \times J$. This is due to the fact that $G_{i,i}(t, t^+) \ne G_{i,i}(t, t^-)$, for all $i \in \{1, \dots, n\}$.

Since for $i \ne j$ the function $G_{i,j}$ can be continuously extended to the diagonal, to avoid tedious notations, in most of the situations, the function G is defined in the whole square $J \times J$, under the assumption that when $t = s$ and $i = j$ we must define $G_{i,i}(t, t)$ as $G_{i,i}(t, t^+)$ or $G_{i,i}(t, t^-)$. Anyway, the qualitative properties as (among others) the regularity, the ess sup, the ess inf, the sign, or the symmetry are not affected with this choice. The same occurs with the different norms of G or with the expression $\int_a^b G(t, s)\, f(s)\, ds$ for any arbitrary function $f \in \mathcal{L}^1(J, \mathbb{R}^n)$.

Remark 1.2.14. We point out that from expression (1.2.14), it is immediate to verify that

$$y(t) = \int_a^b G(t, s)\, f(s)\, ds$$

is the unique solution of problem

$$y'(t) = A(t)\, y(t) + f(t), \quad t \in J, \quad B\, y(a) + C\, y(b) = 0.$$

Moreover,

$$z(t) = \phi(t)\, M_\phi^{-1}\, h$$

is the unique solution of problem

$$z'(t) = A(t)\, z(t), \quad t \in J, \quad B\, z(a) + C\, z(b) = h.$$

Due to the regularity of the matrix function A, we have that function ϕ is an absolutely continuous function. In consequence G is continuous on the triangles $\{(t, s) \in \mathbb{R}^2, \quad a \le s < t \le b\}$ and $\{(t, s) \in \mathbb{R}^2, \quad a \le t < s \le b\}$.

However, along the diagonal, we have that

$$G(t^+, t) = G(t, t^-) = -\phi(t) \, M_\phi^{-1} \, C \, \phi(b) \, \phi^{-1}(t) + I_n$$

and

$$G(t^-, t) = G(t, t^+) = -\phi(t) \, M_\phi^{-1} \, C \, \phi(b) \, \phi^{-1}(t).$$

Therefore, the diagonal elements of the matrix function

$$G(t, s) \equiv (G_{i,j}(t, s))_{i, j \in \{1, \dots, n\}}$$

have a jump (which equals the identity matrix) on the diagonal of $J \times J$.

In others words, for all $i \in \{1, \dots, n\}$, the following equalities hold:

$$\lim_{s \to t^+} G_{i,i}(s, t) = \lim_{s \to t^-} G_{i,i}(t, s) = 1 + \lim_{s \to t^+} G_{i,i}(t, s) = 1 + \lim_{s \to t^-} G_{i,i}(s, t).$$

$$(1.2.16)$$

Now we will verify that for any $s \in (a, b)$ fixed, function $G(\cdot, s) : J \setminus \{s\} \longrightarrow \mathbb{R}^n$ is a solution of the homogeneous system (1.2.3). The result is the following.

Lemma 1.2.15. *For all $s \in (a, b)$ given, let us denote for all $t \in J$, $f_s(t) \equiv G(t, s)$. Then function f_s satisfies the following properties:*

(h1) $f_s \in \mathscr{AC}([s, b], \mathbb{R}^n) \cup \mathscr{AC}([a, s], \mathbb{R}^n)$ *and*

$$f_s'(t) = A(t) \, f_s(t), \quad \text{for a. e. } t \in J \setminus \{s\}.$$

(h2) $B \, f_s(a) + C \, f_s(b) = 0.$
(h3) $\displaystyle\lim_{t \to s^+} (f_s(t))_{i,i} = 1 + \lim_{t \to s^-} (f_s(t))_{i,i} \quad$ *for all $i \in \{1, \dots, n\}$.*

Proof. First, notice that since function ϕ is an absolutely continuous function on J, the expression (1.2.15) implies that for all $s \in J$, $f_s \in \mathscr{AC}([s, b], \mathbb{R}^n)$. Moreover, by using the definition of fundamental matrix, we deduce that for all $t \in (s, b)$, the following equality holds:

$$f_s'(t) = -\phi'(t) \, M_\phi^{-1} \, C \, \phi(b) \, \phi^{-1}(s) + \phi'(t) \, \phi^{-1}(s)$$

$$= A(t) \left(-\phi(t) \, M_\phi^{-1} \, C \, \phi(b) \, \phi^{-1}(s) + \phi(t) \, \phi^{-1}(s) \right) = A(t) \, f_s(t).$$

Analogously, for all $t \in (a, s)$, we have that $f_s \in \mathscr{AC}([a, s], \mathbb{R}^n)$ and

$$f_s'(t) = -\phi'(t) \, M_\phi^{-1} \, C \, \phi(b) \, \phi^{-1}(s) = A(t) \, f_s(t),$$

and condition *(h1)* is fulfilled.

To verify condition $(h2)$, we have that

$$B f_s(a) + C f_s(b) = -B \phi(a) M_\phi^{-1} C \phi(b) \phi^{-1}(s)$$
$$-C \phi(b) M_\phi^{-1} C \phi(b) \phi^{-1}(s) + C \phi(b) \phi^{-1}(s)$$
$$= \left(-B \phi(a) M_\phi^{-1} - C \phi(b) M_\phi^{-1} + I_n\right) C \phi(b) \phi^{-1}(s)$$
$$= \left(-M_\phi M_\phi^{-1} + I_n\right) C \phi(b) \phi^{-1}(s) = 0.$$

Condition $(h3)$ is a direct consequence of (1.2.16). □

Properties $(h1)$–$(h3)$ are known in the literature as the axiomatic definition of a Green's function.

These properties allow us to introduce the concept of Green's function related to the first-order linear system (1.2.3) as follows.

Definition 1.2.16. We say that G is a *Green's function* for problem (1.2.3) if it satisfies the following properties:

(G1) $G \equiv (G_{i,j})_{i,j \in \{1,...,n\}} : (J \times J) \backslash \{(t,t), t \in J\} \to \mathcal{M}_{n \times n}$.
(G2) G is absolutely continuous on the triangles $\{(t,s) \in \mathbb{R}^2, \quad a \le s < t \le b\}$ and $\{(t,s) \in \mathbb{R}^2, \quad a \le t < s \le b\}$.
(G3) For all $i \ne j$ the scalar functions $G_{i,j}$ have a continuous extension to $J \times J$.
(G4) For all $s \in (a,b)$, the following inequality holds:

$$\frac{\partial G}{\partial t}(t,s) = A(t) G(t,s), \quad \text{for a. e. } t \in J \backslash \{s\}.$$

(G5) For all $s \in (a,b)$ and $i \in \{1, \ldots, n\}$, the following equalities are fulfilled:

$$\lim_{s \to t^+} G_{i,i}(s,t) = \lim_{s \to t^-} G_{i,i}(t,s) = 1 + \lim_{s \to t^+} G_{i,i}(t,s) = 1 + \lim_{s \to t^-} G_{i,i}(s,t).$$

(G6) For each $s \in (a,b)$, the function $t \to G(t,s)$ satisfies the boundary conditions

$$B G(a,s) + C G(b,s) = 0.$$

Now, we are in a position to prove the following existence and uniqueness result of a Green's function.

Theorem 1.2.17. *Problem (1.2.3) has only the trivial solution if and only if there exists a unique Green's function related to this problem.*

Proof. First, assume that $u = 0$ is the unique solution of problem (1.2.3). From Theorem 1.2.3 and Corollary 1.2.4, we know that problem (1.2.3) has only the trivial solution if and only if M_ϕ is invertible. In such a case, the function G given by expression (1.2.15) is well defined and it satisfies properties (G1)–(G6).

To see that such Green's function is unique, let $H : (J \times J)\backslash\{(t,t), \ t \in J\} \longrightarrow \mathcal{M}_{n \times n}$ be a function that satisfies properties (G1)–(G6) too, and define $y : J \longrightarrow \mathbb{R}^n$ as

$$y(t) = \int_a^b H(t,s) \, f(s) \, ds = \int_a^t H(t,s) \, f(s) \, ds + \int_t^b H(t,s) \, f(s) \, ds.$$

From conditions (G4) and (G5) we deduce that

$$y'(t) = \int_a^t \frac{\partial H}{\partial t}(t,s) \, f(s) \, ds + \int_t^b \frac{\partial H}{\partial t}(t,s) \, f(s) \, ds + H(t,t^-) \, f(t) - H(t,t^+) \, f(t)$$

$$= \int_a^t A(t) \, H(t,s) \, f(s) \, ds + \int_t^b A(t) \, H(t,s) \, f(s) \, ds + f(t)$$

$$= A(t) \int_a^b H(t,s) \, f(s) \, ds + f(t)$$

$$= A(t) \, y(t) + f(t), \quad \text{for a. e. } t \in J.$$

Moreover, condition (G6) implies that

$$B \, y(a) + C \, y(b) = B \int_a^b H(a,s) \, f(s) \, ds + C \int_a^b H(b,s) \, f(s) \, ds$$

$$= \int_a^b (B \, H(a,s) + C \, H(b,s)) \, f(s) \, ds = 0.$$

In consequence function y is a solution of problem (1.2.1)–(1.2.2) and

$$(x - y)(t) = \int_a^b (G(t,s) - H(t,s)) \, f(s) \, ds$$

is a solution of (1.2.3).

Since system (1.2.3) has only the trivial solution, we deduce that

$$\int_a^b (G(t,s) - H(t,s)) \, f(s) \, ds = 0 \quad \text{for all } f \in \mathcal{L}^1(J, \mathbb{R}^n),$$

which implies that $G \equiv H$ on $(J \times J)\backslash\{(t,t), \ t \in J\}$.

Reciprocally, let G be the unique Green's function related to problem (1.2.3).

Suppose that there is $\Phi \in \mathcal{AC}(J, \mathbb{R}^n)$, a nontrivial solution of the homogeneous problem (1.2.3). In this case, it is immediate to verify that for every $\lambda \in \mathbb{R}$

$$H_\lambda(t,s) := G(t,s) + \Phi(t)\lambda, \quad t, s \in J,$$

is a family of Green's functions related to problem (1.2.3).

In consequence, due to the uniqueness of Green's function, we conclude that $\Phi = 0$, i.e., the trivial solution is the unique solution of the homogeneous system (1.2.3).   □

Notice that if problem (1.2.3) has nontrivial solutions, since M_ϕ^{-1} is not defined, the construction of Green's function given along this section fails, so we cannot guarantee the existence of such a function. This case is pointed out in the following example.

Example 1.2.18. Consider the problem (1.2.1)–(1.2.2), with $J = [0, 2\pi]$,

$$A(t) = \begin{pmatrix} 0 & 1 \\ -1 & 0 \end{pmatrix}, \quad B = \begin{pmatrix} 1 & 0 \\ 0 & 1 \end{pmatrix}, \quad C = \begin{pmatrix} -1 & 0 \\ 0 & -1 \end{pmatrix} \text{ and } h = \begin{pmatrix} 0 \\ 0 \end{pmatrix}.$$

It is obvious that the fundamental matrix $\phi(t) = \begin{pmatrix} \cos t & \sin t \\ -\sin t & \cos t \end{pmatrix}$ is a solution of

the homogeneous problem (1.2.3) and $M_\phi = \begin{pmatrix} 0 & 0 \\ 0 & 0 \end{pmatrix}$. In particular $u = 0$ is not its

unique solution

As we have seen in the proof of Theorem 1.2.17 if there is $G : (J \times J)\setminus\{(t, t), t \in J\} \longrightarrow \mathcal{M}_{2\times2}$, a Green's function related to this system, then conditions (G1)–(G6) imply that for all $f \in \mathcal{L}^1(J, \mathbb{R}^2)$, the function

$$x(t) = \int_0^{2\pi} G(t, s)\, f(s)\, ds, \quad t \in J,$$

is a solution of our problem.

This property is equivalent to say that operator L, defined in (1.2.4), is surjective. But since rank $(B \mid C) = 2$, Theorem 1.2.10(v) tell us that, in this case, M_ϕ is invertible, which is a contradiction.

Surprisingly, it is possible to construct Green's functions, even in the case of the homogeneous problem (1.2.3) has nontrivial solutions. This is the case in which operator L is surjective and the condition of linear independence (1.2.11) is not fulfilled.

To see this, consider the following example.

Example 1.2.19. Consider the problem (1.2.1)–(1.2.2), with $J = [0, 2\pi]$,

$$A(t) = \begin{pmatrix} 0 & 1 \\ -1 & 0 \end{pmatrix}, \quad B = \begin{pmatrix} 1 & 0 \\ 0 & 0 \end{pmatrix}, \quad C = \begin{pmatrix} 1 & 0 \\ 0 & 0 \end{pmatrix} \text{ and } h = \begin{pmatrix} 0 \\ 0 \end{pmatrix}.$$

First, we note that the matrix $M(t) = \begin{pmatrix} \sin t & \sin t \\ \cos t & \cos t \end{pmatrix}$ is a nontrivial solution of the

homogeneous problem (1.2.3).

Moreover, $\phi(t) = \begin{pmatrix} \cos t & \sin t \\ -\sin t & \cos t \end{pmatrix}$, $M_\phi = \begin{pmatrix} 2 & 0 \\ 0 & 0 \end{pmatrix}$, and rank $(B \mid C) = 1$.

Now, from expression (1.2.8), we can obtain as many Green's functions as $\lambda \in \mathbb{R}^2$ solve system (1.2.9) (with $h = 0$).

Notice that (1.2.9) can be rewritten in this situation as

$$\begin{pmatrix} 2 & 0 \\ 0 & 0 \end{pmatrix} \begin{pmatrix} \lambda_1 \\ \lambda_2 \end{pmatrix} = -\begin{pmatrix} 1 & 0 \\ 0 & 0 \end{pmatrix} \begin{pmatrix} 1 & 0 \\ 0 & 1 \end{pmatrix} \int_0^{2\pi} \begin{pmatrix} \cos s & -\sin s \\ \sin s & \cos s \end{pmatrix} \begin{pmatrix} f_1(s) \\ f_2(s) \end{pmatrix} ds.$$

It is immediate to verify that this system is compatible for all $f \in \mathscr{L}^1(J, \mathbb{R}^2)$ and the solutions of this system are given by the following expression:

$$\lambda_1 = \frac{1}{2} \int_0^{2\pi} (-f_1(s) \cos s + f_2(s) \sin s) \, ds, \quad \lambda_2 \in \mathbb{R}.$$

Substituting this last expression into (1.2.8), we define, for every $\lambda_2 \in \mathbb{R}$, the function

$$G(t, s) \equiv (G_{i,j}(t, s))_{i,j=1,2} : J \times J \backslash \{(t, t), \, t \in J\} \longrightarrow \mathscr{M}_{2\times 2},$$

as follows:

$$G_{1,1}(t, s) = \begin{cases} -\dfrac{1}{2} \cos t \cos s + \cos(t - s) + \lambda_2 \sin t, & \text{if} \quad 0 \le s < t \le 2\pi, \\[3mm] -\dfrac{1}{2} \cos t \cos s + \lambda_2 \sin t, & \text{if} \quad 0 \le t < s \le 2\pi, \end{cases}$$

$$G_{1,2}(t, s) = \begin{cases} \dfrac{1}{2} \cos t \sin s + \sin(t - s) + \lambda_2 \sin t, & \text{if} \quad 0 \le s < t \le 2\pi, \\[3mm] \dfrac{1}{2} \cos t \sin s + \lambda_2 \sin t, & \text{if} \quad 0 \le t < s \le 2\pi, \end{cases}$$

$$G_{2,1}(t, s) = \begin{cases} \dfrac{1}{2} \sin t \cos s - \sin(t - s) + \lambda_2 \cos t, & \text{if} \quad 0 \le s < t \le 2\pi, \\[3mm] \dfrac{1}{2} \sin t \cos s + \lambda_2 \cos t, & \text{if} \quad 0 \le t < s \le 2\pi \end{cases}$$

$$G_{2,2}(t, s) = \begin{cases} -\dfrac{1}{2} \sin t \sin s + \cos(t - s) + \lambda_2 \cos t, & \text{if} \quad 0 \le s < t \le 2\pi, \\[3mm] -\dfrac{1}{2} \sin t \sin s + \lambda_2 \cos t, & \text{if} \quad 0 \le t < s \le 2\pi. \end{cases}$$

Next we verify that the function G is, for any $\lambda_2 \in \mathbb{R}$, a Green's function related to the considered problem.

Clearly, for all $s \in (0, 2\pi)$, is fulfilled that

$$\frac{\partial G}{\partial t}(t, s) = A(t)\, G(t, s) \qquad \text{for all } [0, 2\pi] \setminus \{s\}.$$

Moreover, for all $s \in (0, 2\pi)$,

$$B\, G(0, s) + C\, G(2\pi, s) = \begin{pmatrix} 1 & 0 \\ 0 & 0 \end{pmatrix} \begin{pmatrix} -\dfrac{\cos s}{2} & \dfrac{\sin s}{2} \\ \lambda_2 & \lambda_2 \end{pmatrix} + \begin{pmatrix} 1 & 0 \\ 0 & 0 \end{pmatrix} \begin{pmatrix} \dfrac{\cos s}{2} & -\dfrac{\sin s}{2} \\ \lambda_2 + \sin s & \lambda_2 + \cos s \end{pmatrix}$$

$$= \begin{pmatrix} 0 & 0 \\ 0 & 0 \end{pmatrix}.$$

Finally, it is immediate to verify that for all $t \in (0, 2\pi)$ the following equality is fulfilled:

$$G(t^+, t) - G(t^-, t) = G(t, t^-) - G(t, t^+) = \begin{pmatrix} 1 & 0 \\ 0 & 1 \end{pmatrix}.$$

So, this problem has an infinity number of Green's functions. Moreover, as we have pointed out in the proof of Theorem 1.2.17, subtracting two functions of this type, we have a multiple of the matrix M, a nontrivial solution of the homogeneous problem (1.2.3).

This example shows that, contrary to Example 1.2.18, it is possible to find Green's functions related to some linear problems that have not a unique solution. Therefore the condition imposed in Theorem 1.2.17 cannot be avoided. In particular, the existence of a Green's function does not imply, by itself, the uniqueness of solutions of the related problem.

With just an overview one could deduce that the difference between the two previous situations is, instead of condition (1.2.11), rank (M_ϕ). In the first example it is equal to zero and in the second situation its value is equal to one. However, this is a false trail. In the following example we present a two-dimensional problem with rank $(M_\phi) = 1$ that has no Green's function.

Example 1.2.20. Consider the problem (1.2.1)–(1.2.2), with $J = [0, 2\pi]$,

$$A(t) = \begin{pmatrix} 0 & 1 \\ -1 & 0 \end{pmatrix}, \quad B = \begin{pmatrix} 0 & 1 \\ 0 & 0 \end{pmatrix}, \quad C = \begin{pmatrix} 0 & 0 \\ 0 & 1 \end{pmatrix} \text{ and } h = \begin{pmatrix} 0 \\ 0 \end{pmatrix}.$$

In this case $M(t) = \begin{pmatrix} \cos t & \cos t \\ -\sin t & -\sin t \end{pmatrix}$ is a nontrivial solution of the homogeneous problem (1.2.3) and rank $(B \mid C) = 2$.

Now, system (1.2.9) (with $h = 0$) is given by

$$\begin{pmatrix} 0 & 1 \\ 0 & 1 \end{pmatrix} \begin{pmatrix} \lambda_1 \\ \lambda_2 \end{pmatrix} = -\begin{pmatrix} 0 & 0 \\ 0 & 1 \end{pmatrix} \begin{pmatrix} 1 & 0 \\ 0 & 1 \end{pmatrix} \int_0^{2\pi} \begin{pmatrix} \cos s & -\sin s \\ \sin s & \cos s \end{pmatrix} \begin{pmatrix} f_1(s) \\ f_2(s) \end{pmatrix} ds.$$

So, the above system is compatible if and only if

$$\int_0^{2\pi} (f_1(s) \sin s + f_2(s) \cos s)\, ds = 0.$$

As a consequence, we have that operator L is not surjective. Arguing as in Example 1.2.18, we deduce that there is no Green's function related to this problem.

In fact the argument used in Examples 1.2.18 and 1.2.20 is valid for any n-dimensional linear system (1.2.3). Indeed, if rank $(B \mid C) = n$ and there is a related Green's function, we have that problem (1.2.1)–(1.2.2) (with $h = 0$) is always solvable for any $f \in \mathscr{L}^1(J, \mathbb{R}^n)$. This implies that operator L defined in (1.2.4) is surjective. As a consequence, Theorem 1.2.10 ensures that problem (1.2.1)–(1.2.2) has a unique solution. This property can be enunciated as follows.

Lemma 1.2.21. *Problem (1.2.1)–(1.2.2) has a unique solution if and only if* rank$(B \mid C) = n$ *and there exists a Green's function related to problem (1.2.3).*

Returning to the expression (1.2.15), it is not difficult to verify that if $A(t)$ and $\int_a^t A(s)\, ds$ commute, then $\phi(t) = \exp\left(\int_a^t A(s)\, ds\right)$ is a fundamental matrix of the linear system (1.2.3). It is obvious that in this case $\phi^{-1}(t) = \exp\left(-\int_a^t A(s)\, ds\right)$.

So, by assuming this standard commutative property, we rewrite expression (1.2.15) as follows:

$$G(t,s) = \begin{cases} e^{\int_a^t A(r)\, dr} \left(B + C\, e^{\int_a^b A(r)\, dr}\right)^{-1} B\, e^{-\int_a^s A(r)\, dr}, & a \le s < t \le b, \\[2ex] -e^{\int_a^t A(r)\, dr} \left(B + C\, e^{\int_a^b A(r)\, dr}\right)^{-1} C\, e^{\int_s^b A(r)\, dr}, & a \le t < s \le b. \end{cases}$$

Now, if the boundary matrices B and C commute with $\exp\left(\int_a^t A(s)\, ds\right)$ for all $t \in J$, then the last expression is given by

$$G(t,s) = \begin{cases} e^{\int_s^t A(r)\, dr} \left(B + C\, e^{\int_a^b A(r)\, dr}\right)^{-1} B, & a \le s < t \le b, \\[2ex] -e^{\int_s^t A(r)\, dr} \left(B + C\, e^{\int_a^b A(r)\, dr}\right)^{-1} C\, e^{\int_a^b A(r)\, dr}, & a \le t < s \le b. \end{cases}$$

This last expression can be applied to the initial value problem (with $B = I_n$ and $C = 0$):

$$G(t,s) = \begin{cases} e^{\int_s^t A(r)\,dr}, & a \le s < t \le b, \\ 0, & a \le t < s \le b, \end{cases}$$

and for the terminal one, $B = 0$ and $C = I_n$,

$$G(t,s) = \begin{cases} 0, & a \le s < t \le b, \\ -e^{\int_s^t A(r)\,dr}, & a \le t < s \le b. \end{cases}$$

Moreover, it remains valid for the periodic case $B = I_n$ and $C = -I_n$

$$G(t,s) = \begin{cases} e^{\int_s^t A(r)\,dr}\left(I_n - e^{\int_a^b A(r)\,dr}\right)^{-1}, & a \le s < t \le b, \\ e^{\int_s^t A(r)\,dr}\left(I_n - e^{\int_a^b A(r)\,dr}\right)^{-1} e^{\int_a^b A(r)\,dr}, & a \le t < s \le b. \end{cases}$$

By using the previous expressions and defining

$$G(a,a) = \lim_{t\to a^+} G(t,a) \quad \text{and} \quad G(b,b) = \lim_{t\to b^-} G(t,b),$$

it is not difficult to verify that if we refer to the initial value problem, function $G(\cdot,a) : J \to \mathbb{R}^n$ is the unique solution of the problem

$$R'(t) = A(t)\,R(t), \quad t \in \mathbb{R}, \quad R(a) = I_n.$$

Concerning the terminal value problem, $G(\cdot,b) : J \to \mathbb{R}^n$ is the unique solution of the problem

$$R'(t) = A(t)\,R(t), \quad t \in \mathbb{R}, \quad R(b) = -I_n.$$

If we refer to the periodic case, $G(\cdot,a) : J \to \mathbb{R}^n$ is the unique solution of the problem

$$R'(t) = A(t)\,R(t), \quad t \in \mathbb{R}, \quad R(a) - R(b) = I_n.$$

We pay now special attention to the case in which the coefficient matrix $A(t)$ is constant. In this case, $\int_a^t A(s)\,ds = A(t-a)$ and this matrix commutes with A.

If we refer to the initial value problem, from the corresponding expression shown above, Green's function satisfies

$$G(t,s) = G(t-s+a,a), \quad \text{if } a \le s \le t \le b, \text{ and } G(t,s) = 0 \text{ if } a \le t < s \le b.$$

When we consider the terminal value problem, Green's function satisfies

$$G(t, s) = G(b + t - s, b), \quad \text{if } a \leq t \leq s \leq b, \text{ and } G(t, s) = 0 \text{ if } a \leq s < t \leq b.$$

Finally, taking into account the periodic case, we have that

$$G(t, s) = G(a + t - s, a), \quad \text{if } a \leq s \leq t \leq b,$$

and

$$G(t, s) = G(b + t - s, a), \quad \text{if } a \leq t < s \leq b.$$

Notice that the functions related to these three problems are constant over the straight lines of slope equals to one. In consequence, in such a cases, it is enough to calculate the value of Green's function on one side of the square of definition and extend it to the whole square $J \times J$. Moreover, the expression of Green's function on each side of the square follows by solving a simple linear homogeneous system. We remark that if the coefficient matrix is not constant, then the previous properties do not hold.

Remark 1.2.22. It is important to mention that although the calculations in expression (1.2.15) are far to be trivial, when the coefficients matrix A is constant, it is not difficult to make an easy code in Mathematica in which such expression can be given explicitly. Of course its applicability depends on the difficulty of the computer in obtaining the exponential matrix of $A(t - a)$; this makes it not valid in every situation.

Next, a code is presented for a particular two-dimensional system. For the general case, it is enough to modify the values of the extremes of the interval $[a, b]$ and those of the constant matrices A, B, and C, by adapting them to each particular situation and dimension.

$a = 0$

$b = 1/2$

$AA = \{\{-2\lambda/3, 0\}\}, \{\{0, -4\lambda/3\}\}$

$BB = \{\{1, 2\}, \{0, 1\}\}$

$CC = \{\{0, -1\}, \{-1, 0\}\}$

Phi[t_-] = MatrixExp[$AA(t - a)$]

MPhi = Inverse[$BB + CC$.Phi[b]]

G1[t_-, s_-] = Phi[$t - s$].MPhi.BB

G2[t_-, s_-] = $-$ Phi[$t - s$].MPhi.CC.Phi[b]

1.2.1 Multipoint Boundary Value Problems

In the previous part of this section we have considered the two-point boundary problem (1.2.1)–(1.2.2). However, in many situations, the system (1.2.1) is subject to are evaluated not only at the extremes of the bounded interval J, but also depend on the value at some inner points.

To be concise, the multipoint boundary conditions follow an expression of this type:

$$\sum_{i=0}^{m} C_i \, x(a_i) = h. \tag{1.2.17}$$

Here m is a positive integer, $C_i \in \mathcal{M}_{n \times n}$ for all $i \in \{0, \ldots, m\}$, $h \in \mathbb{R}^n$ and $a = a_0 < a_1 < \cdots < a_m = b$.

In this case, the existence and uniqueness of solutions of problem (1.2.1), (1.2.17) can be deduced from a direct translation to a two-point $(m-1) \times n$-dimensional first-order boundary value problem. This property is illustrated in the following result.

Lemma 1.2.23. Let $\bar{a} = a + \frac{(b-a)}{m}$. Define, for every $j = 0, \ldots, m-1$ and $t \in J_1 = [a, \bar{a}]$, $t_j = a_j + m \, (t-a) \frac{a_{j+1} - a_j}{b-a}$. Then $x : J \longrightarrow \mathbb{R}^n$ is a solution of the multipoint problem (1.2.1), (1.2.17) if and only if

$$\bar{x} \equiv (\bar{x}_1, \ldots, \bar{x}_m)^T : J_1 \longrightarrow \mathbb{R}^{n \times (m-1)},$$

defined as $\bar{x}_j(t) = x\left(t_{j-1}\right)$ for all $j = 1, \ldots, m$ and $t \in J_1$, is a solution of problem

$$\begin{cases} \bar{x}'(t) = \bar{A}(t) \, \bar{x}(t) + \bar{f}(t), & t \in J_1 \\[2mm] \bar{B} \, \bar{x}(a) + \bar{C} \, \bar{x}(\bar{a}) = \bar{h}, \end{cases} \tag{1.2.18}$$

where

$$\bar{A}(t) = \frac{m}{b-a} \begin{pmatrix} (a_1 - a) \, A\,(t_0) & 0 & \cdots & 0 \\ 0 & (a_2 - a_1) \, A\,(t_1) & \cdots & 0 \\ \vdots & \vdots & \ddots & \vdots \\ 0 & 0 & \cdots & (b - a_{m-1}) \, A\,(t_{m-1}) \end{pmatrix},$$

$$\bar{f}(t) = \frac{m}{b-a} \begin{pmatrix} (a_1 - a) \, f\,(t_0) \\ (a_2 - a_1) \, f\,(t_1) \\ \vdots \\ (b - a_{m-1}) \, f\,(t_{m-1}) \end{pmatrix},$$

$$\bar{B} = \left(\begin{array}{c|ccc} C_0 & C_1 & \cdots & C_{m-1} \\ \hline 0 & & I_{n \times (m-1)} & \end{array} \right),$$

$$\bar{C} = \left(\begin{array}{ccc|c} 0 & \cdots & 0 & C_n \\ \hline & -I_{n \times (m-1)} & & 0 \end{array} \right)$$

and

$$\bar{h} = \left(\begin{array}{c} h \\ 0 \\ \vdots \\ 0 \end{array} \right).$$

Proof. The proof follows as a straightforward change of variables and by taking into account that x must be a continuous function at points a_i, $i = 1, \ldots, m - 1$. □

Remark 1.2.24. Notice that the matrix function $\bar{G} : J_1 \times J_1 \longrightarrow \mathcal{M}_{(n \times m) \times (n \times m)}$ related to the two-point boundary value problem (1.2.18) gives the values of the function $\bar{x}$ on J_1. Once we have obtained the expression of the vector $\bar{x}$, then for any $t \in J$, we know that there is $j \in \{0, \ldots, m - 1\}$ such that $t \in [a_j, a_{j+1}]$. In such a case, we deduce that

$$x(t) = \bar{x}_{j-1} \left(a + \frac{(t - a_j)(b - a)}{m(a_{j+1} - a_j)} \right). \tag{1.2.19}$$

Notice that if $f \in \mathcal{L}^1(J, \mathbb{R}^n)$ and $A \in \mathcal{L}^1(J, \mathcal{M}_{n \times n})$, then $\bar{f} \in \mathcal{L}^1(J, \mathbb{R}^{n \times m})$ and $\bar{A} \in \mathcal{L}^1(J, \mathcal{M}_{(n \times m) \times (n \times m)})$. The regularity is preserved even in the continuous case, but it can be lost if we are speaking about more regular functions.

As a consequence of the regularity property and the previous lemma, we deduce the following existence and uniqueness result for problem (1.2.1), (1.2.17).

Corollary 1.2.25. *Let $A \in \mathcal{L}^1(J, \mathcal{M}_{n \times n})$ and $f \in \mathcal{L}^1(J, \mathbb{R}^n)$, $C_i \in \mathcal{M}_{n \times n}$ for all $i \in \{0, \ldots, m\}$ and $h \in \mathbb{R}^n$.*
Let

$$\bar{M} = \bar{B} \bar{\phi}(a) + \bar{C} \bar{\phi}(\bar{a}),$$

and $\bar{\phi}$ a fundamental matrix of the system

$$\bar{\phi}'(t) = \bar{A}(t) \bar{\phi}(t), \qquad t \in [a, \bar{a}],$$

with $\bar{C}$, $\bar{D}$, and $\bar{A}(t)$ defined in the enunciate of Lemma 1.2.23.
Then problem (1.2.1), (1.2.17) has a unique solution if and only if $\det \bar{M} \neq 0$.

From this property, using the results given in the previous part of this section for two-point boundary value problems, it is possible to deduce the existence and the expression of the related Green's function together with its properties for this new situation.

Example 1.2.26. Consider, for any $\lambda \in \mathbb{R}$, the multipoint first- order linear problem:

$$x'(t) + \lambda \, x(t) = t, \quad t \in [0, 1], \qquad x(0) + 2 \, x(1/3) - x(1) = 0.$$

In this situation, $n = 1$, $a = 0$, $a_1 = 1/3$, $b = 1$, $m = 2$, $C_0 = 1$, $C_1 = 2$ and $C_2 = -1$. Moreover, the matrices and vectors defined as in Lemma 1.2.23 are given by the following expressions:

$$\bar{x}(t) = \begin{pmatrix} x\left(\frac{2t}{3}\right) \\ x\left(\frac{1+4t}{3}\right) \end{pmatrix}, \quad \bar{A}(t) = -\begin{pmatrix} \frac{2\lambda}{3} & 0 \\ 0 & \frac{4\lambda}{3} \end{pmatrix}, \quad \bar{f}(t) = \begin{pmatrix} \frac{4t}{9} \\ \frac{4+8t}{9} \end{pmatrix},$$

$$\bar{B} = \begin{pmatrix} 1 & 2 \\ 0 & 1 \end{pmatrix}, \quad \bar{C} = \begin{pmatrix} 0 & -1 \\ -1 & 0 \end{pmatrix}, \quad \text{and} \quad \bar{h} = \begin{pmatrix} 0 \\ 0 \end{pmatrix}.$$

To evaluate the expression (1.2.15) for this particular case, we make use of the Mathematica program shown in Remark 1.2.22 and we arrive at the following expression:

$$G(t,s) = \begin{pmatrix} \frac{e^{\frac{1}{3}(2s-2t+3)\lambda}}{-1+2e^{2\lambda/3}+e^{\lambda}} & \frac{e^{\frac{1}{3}(4s-2t+1)\lambda}}{-1+2e^{2\lambda/3}+e^{\lambda}} \\ \frac{e^{\frac{2}{3}(s-2t+1)\lambda}}{-1+2e^{2\lambda/3}+e^{\lambda}} & \frac{e^{\frac{1}{3}(2s-2t+1)\lambda}\left(2+e^{\lambda/3}\right)}{-1+2e^{2\lambda/3}+e^{\lambda}} \end{pmatrix}, \quad \text{if } 0 \le s < t \le 1/2,$$

and

$$G(t,s) = \begin{pmatrix} \frac{e^{\frac{2}{3}(s-t)\lambda}\left(1-2e^{2\lambda/3}\right)}{-1+2e^{2\lambda/3}+e^{\lambda}} & \frac{e^{\frac{1}{3}(4s-2t+1)\lambda}}{-1+2e^{2\lambda/3}+e^{\lambda}} \\ \frac{e^{\frac{2}{3}(s-2t+1)\lambda}}{-1+2e^{2\lambda/3}+e^{\lambda}} & \frac{e^{\frac{4}{3}(s-t)\lambda}}{-1+2e^{2\lambda/3}+e^{\lambda}} \end{pmatrix}, \quad \text{if } 0 \le t < s \le 1/2.$$

So, since

$$\bar{M} = \begin{pmatrix} 1 & 2 - e^{-2\lambda/3} \\ -e^{-\lambda/3} & 1 \end{pmatrix},$$

we conclude that our problem has a unique solution if and only if $\lambda \ne \log\left(\sqrt{5} - 2\right)$. In such a case, we have that

$$\begin{pmatrix} \bar{x}_1(t) \\ \bar{x}_2(t) \end{pmatrix} = \int_0^{1/2} G(t,s) \, \bar{f}(s) \, ds = \begin{pmatrix} \frac{2\lambda t + \frac{(\lambda+6)e^{\lambda - \frac{2\lambda t}{3}}}{2e^{2\lambda/3}+e^{\lambda}-1} - 3}{3\lambda^2} \\ \frac{\lambda + 4\lambda t + \frac{(\lambda+6)e^{\frac{2}{3}\lambda(1-2t)}}{2e^{2\lambda/3}+e^{\lambda}-1} - 3}{3\lambda^2} \end{pmatrix}.$$

Now, by making use of expression (1.2.19), we conclude that the unique solution that we are looking for is given by the expression

$$x(t) = \frac{3\lambda t + \frac{(\lambda+6)e^{\lambda-\lambda t}}{2e^{2\lambda/3}+e^{\lambda}-1} - 3}{3\lambda^2}, \quad t \in [0, 1].$$

1.3 Green's Functions for Adjoint Operators

This section is devoted to the relationship between Green's function of a linear operator and Green's function of the adjoint operator.

First, let us recall that [55, p. 786], given a Hilbert space H equipped with the inner product $\langle \cdot, \cdot \rangle_H$ and $T : H \longrightarrow H$ a linear and continuous operator, the adjoint operator T^* of the operator T is defined as the unique linear and continuous operator $T^* : H \longrightarrow H$ that satisfies the following equality:

$$\langle T u, v \rangle_H = \langle u, T^* v \rangle_H, \qquad \forall u, v \in H.$$

Moreover, $T^{**} = T$ and $\|T\| = \|T^*\|$.

In our case, by considering a two-variable matrix function $K : J \times J \to \mathcal{M}_{n \times n}$, such that $K \in \mathcal{L}^2(J \times J, \mathcal{M}_{n \times n})$, we can define the integral operator $T : \mathcal{L}^2(J, \mathbb{R}^n) \to \mathcal{L}^2(J, \mathbb{R}^n)$ as follows:

$$T u(t) = \int_a^b K(t, s) u(s) \, ds, \quad t \in J. \tag{1.3.1}$$

It is not difficult to verify that such operator is well defined (see [55, p. 786] for details) and that $\mathcal{L}^2(J, \mathbb{R}^n)$ is a Hilbert space with the inner product

$$\langle u, v \rangle_2 := \int_a^b \langle u(t), v(t) \rangle \, dt,$$

where $\langle \cdot, \cdot \rangle$ denotes the usual scalar product in $\mathbb{R}^n$.

We are interested in calculating the expression of the adjoint operator of T. So we arrive at the following result.

Theorem 1.3.1. *Let T be given from expression (1.3.1) and denote by K^T the transpose matrix kernel of K. Then, for all $v \in H$, the adjoint operator T^* is given by the following expression:*

$$T^* v(t) = \int_a^b K^T(s, t) v(s) \, ds, \quad t \in J. \tag{1.3.2}$$

Proof. To prove this result, we deduce, from Fubini's theorem, the following identity:

$$\langle T\,u, v\rangle_2 = \int_a^b \left\langle \int_a^b K(t,s)\,u(s)\,ds, v(t)\right\rangle dt$$

$$= \int_a^b \sum_{i=1}^n \left(\sum_{j=1}^n \int_a^b K_{i,j}(t,s)\,u_j(s)\,ds\right) v_i(t)\,dt$$

$$= \int_a^b \sum_{i=1}^n \left(\sum_{j=1}^n \int_a^b K_{i,j}(t,s)\,u_j(s)\,v_i(t)\,ds\right) dt$$

$$= \int_a^b \sum_{j=1}^n \left(\sum_{i=1}^n \int_a^b K_{i,j}(t,s)\,v_i(t)\,dt\right) u_j(s)\,ds$$

$$= \int_a^b \sum_{j=1}^n \left(\sum_{i=1}^n \int_a^b K_{i,j}(s,t)\,v_i(s)\,ds\right) u_j(t)\,dt$$

$$= \int_a^b \left\langle u(t), \int_a^b K^T(s,t)\,v(s)\,ds\right\rangle dt$$

$$= \left\langle u, \int_a^b K^T(s,t)\,v(s)\,ds\right\rangle_2 .$$

From the definition of the adjoint operator, due its uniqueness, we conclude that the adjoint operator T^* of the integral operator T is given by expression (1.3.2) and the proof is finished.   □

Remark 1.3.2. Operator T is self-adjoint on H, i.e., $T = T^*$, if and only if the following symmetric property holds:

$$K(t,s) = K^T(s,t)\qquad \text{for a.e. } (t,s) \in J \times J. \tag{1.3.3}$$

Now we address our attention to the particular case in which the kernel K, which characterizes the operator T, is Green's function related to the linear system (1.2.1)–(1.2.2) given by expression (1.2.15). By definition of Green's function, we know that for all $u \in \mathscr{L}^2(J, \mathbb{R}^n)$, the function

$$w(t) = T\,u(t) = \int_a^b G(t,s)\,u(s)\,ds,\qquad t \in J,$$

satisfies that $w \in W := \{x \in \mathscr{AC}(J, \mathbb{R}^n),\ x' \in \mathscr{L}^2(J, \mathbb{R}^n)\}$ and, moreover,

$$L\,w(t) := w'(t) - A(t)\,w(t) = u(t),\qquad t \in J,\qquad B\,w(a) + C\,w(b) = 0.$$

So, by denoting $D(L) = \{w \in W : B\,w(a) + C\,w(b) = 0\}$, we have that $L = T^{-1}$ in $D(L)$.

We are interested in knowing what is, if it exists, the differential operator L^* and the suitable space $D(L^*)$, related to the kernel $G^*(t,s) := G^T(s,t)$. That is to say, we are looking for $L^* = (T^*)^{-1}$ in $D(L^*)$. In particular L^* is bijective in $D(L^*)$.

To this end, since condition (G5), in Definition 1.2.16, holds for $-G^*(t,s)$, we will verify that function $-G^*$ satisfies the rest of the properties (G1)–(G6) that characterize Green's function related to a suitable differential operator.

Obviously, conditions (G1)–(G3) are trivially fulfilled. To verify condition (G4), it suffices to use expression (1.2.15) and to take into account that

$$((\phi^{-1}(t))^T)' = \quad ((\phi^{-1}(t))')^T \quad = -(\phi^{-1}(t)\,\phi'(t)\,\phi^{-1}(t))^T$$
$$= -(\phi^{-1}(t)\,A(t))^T = -A^T(t)((\phi^{-1}(t))^T).$$

So, we deduce that for all $s \in (a,b)$

$$\frac{\partial(-G^*)}{\partial t}(t,s) = -A^T(t)\,(-G^*(t,s)), \qquad \text{for a. e. } t \in J \setminus \{s\}.$$

In consequence, given $v \in \mathscr{L}^2(J,\mathbb{R}^n)$, we know that

$$z(t) = T^*\,v(t) := \int_a^b G^*(t,s)\,v(s)\,ds = \int_a^b (-G^*(t,s))\,(-v(s))\,ds$$

if and only if $z \in W$ and

$$z'(t) + A^T(t)\,z(t) = -v(t), \quad t \in J.$$

So we have deduced that the differential adjoint operator is given by

$$L^*\,z(t) := -z'(t) - A^T(t)\,z(t), \quad t \in J.$$

To characterize the space $D(L^*)$ in which the operator L^* is defined, let $u, v \in \mathscr{L}^2(J,\mathbb{R}^n)$ be two arbitrary elements, and $w = T\,u$ and $z = T^*\,v$. As we have seen, $w, z \in W$ and $w' - A(t)\,w = u$ and $-z' - A^T(t)\,z = v$. As a consequence

$$\langle T\,u, v\rangle_2 = \langle u, T^*\,v\rangle_2 = \langle w' - A(\cdot)\,w, z\rangle_2 = \langle w', z\rangle_2 - \langle A(\cdot)\,w, z\rangle_2$$
$$= \langle w(b), z(b)\rangle - \langle w(a), z(a)\rangle - \langle w, z'\rangle_2 - \langle w, A^T(\cdot)\,z\rangle_2$$
$$= \langle w(b), z(b)\rangle - \langle w(a), z(a)\rangle + \langle w, -z' - A^T(\cdot)\,z\rangle_2$$
$$= \langle w(b), z(b)\rangle - \langle w(a), z(a)\rangle + \langle T\,u, v\rangle_2.$$

So we conclude that

$$D(L^*) = \{z \in W;\ \langle w(b), z(b)\rangle = \langle w(a), z(a)\rangle, \quad \text{for all } w \in D(L)\}.$$

This characterization of the space $D(L^*)$ is unsatisfactory. We expect to have a set

$$D(L^*) = \{z \in W, \ D z(a) + E z(b) = 0\},$$

with $D, \ E \in \mathcal{M}_{n \times n}$ two matrices determined by B and C. Although we cannot obtain their exact expression for every couple of matrices B and C, we are able to deduce some necessary properties that such coefficient matrices D and E must satisfy.

The first necessary condition follows from the assumption of the existence of Green's functions for the two differential operators and from Lemma 1.2.21:

$$\text{rank } (D \mid E) = \text{rank } (B \mid C) = n.$$

The second necessary condition is related to the last property (G6) in Definition 1.2.16 that must satisfy $-G^*$, i.e., if $s \in (a, b)$, then

$$0 = D G^*(a, s) + E G^*(b, s) = D G^T(s, a) + E G^T(s, b).$$

Such equality is fulfilled if and only if

$$G(s, a) D^T + G(s, b) E^T = 0.$$

Now, using expression (1.2.15) we have that this last equality holds if and only if

$$\phi(s) \left(-M_\phi^{-1} C \phi(b) \phi^{-1}(a) D^T + \phi^{-1}(a) D^T - M_\phi^{-1} C E^T \right) = 0, \quad s \in (a, b),$$

or, which is the same,

$$C \phi(b) \phi^{-1}(a) D^T - M_\phi \phi^{-1}(a) D^T + C E^T = -B D^T + C E^T = 0.$$

In the particular case of B invertible, we have that

$$B w(a) + C w(b) = 0 \quad \text{if and only if} \quad w(a) = -B^{-1} C w(b).$$

As a consequence, if $w \in D(L)$, we have that

$$\langle w(a), z(a) \rangle = \langle -B^{-1} C w(b), z(a) \rangle = \langle w(b), -(B^{-1} C)^T z(a) \rangle$$

and

$$D(L^*) = \{z \in W; \ (B^{-1} C)^T z(a) + z(b) = 0\}.$$

In an analogous way, we deduce that if there is C^{-1}, then

$$D(L^*) = \{z \in W; \ z(a) + (C^{-1} B)^T z(b) = 0\}.$$

Example 1.3.3. If we consider the periodic boundary condition $w(a) = w(b)$, that corresponds to the choice of $B = I_n$ and $C = -I_n$, we have that

$$\langle w(b), z(b) \rangle = \langle w(a), z(a) \rangle \quad \text{for all } w \in D(L) \quad \text{if and only if} \quad z(a) = z(b).$$

Thus, if we consider the periodic conditions, we conclude that

$$D(L) = D(L^*) = \{x \in W, \quad x(a) = x(b)\}.$$

However, this is not the usual situation; if we deal with the initial value problem $w(a) = 0$ ($B = I_n$ and $C = 0$), we have that

$$\langle w(b), z(b) \rangle = \langle w(a), z(a) \rangle \quad \text{for all } w \in D(L) \quad \text{if and only if} \quad z(b) = 0,$$

that is, for the initial problem, the adjoint operator is related to the terminal problem ($B = 0$ and $C = I_n$) and vice versa.

Example 1.3.4. Consider the first-order scalar periodic boundary value problem

$$u'(t) = a(t)\,u(t) + f(t), \ t \in [0, 2\pi], \quad u(0) = u(2\pi),$$

which is of the form (1.2.1)–(1.2.2) for $J = [0, 2\pi]$, $A(t) \equiv a(t)$, $B = 1$, $C = -1$ and $h = 0$.
    Obviously $\phi(t) = e^{\int_0^t a(s)\,ds}$ and $M_\phi = 1 - e^{\int_0^{2\pi} a(s)\,ds}$.
    So, we have that this problem has a unique solution if and only if

$$\int_0^{2\pi} a(s)\,ds \neq 0.$$

In this case, from (1.2.15), we arrive at the following expression:

$$G(t,s) = \frac{1}{1 - e^{\int_0^{2\pi} a(r)\,dr}} \begin{cases} e^{\int_s^t a(r)\,dr}, & 0 \leq s < t \leq 2\pi, \\ e^{\int_s^t a(r)\,dr}\, e^{\int_0^{2\pi} a(r)\,dr}, & 0 \leq t < s \leq 2\pi. \end{cases}$$

So, from Theorem 1.3.1, we have that the expression of the kernel related to the adjoint operator is given by

$$G^*(t,s) = \frac{1}{1 - e^{\int_0^{2\pi} a(r)\,dr}} \begin{cases} e^{\int_t^s a(r)\,dr}\, e^{\int_0^{2\pi} a(r)\,dr}, & 0 \leq s < t \leq 2\pi, \\ e^{\int_t^s a(r)\,dr}, & 0 \leq t < s \leq 2\pi, \end{cases}$$

which gives us the solutions of the periodic boundary value problem:

$$-u'(t) = a(t)\,u(t) + f(t), \ t \in [0, 2\pi], \quad u(0) = u(2\pi).$$

Obviously this problem is not self-adjoint.

We remark that since it does not satisfy condition (G5) in Definition 1.2.16, G^* is not a Green's function of any linear problem. We can refer to $-G^*$ as Green's function of the operator $u'(\cdot)+a(\cdot)\,u(\cdot)$ defined in the space of absolutely continuous functions coupled with 2π- periodic boundary conditions.

Example 1.3.5. Now we point out that, if we are considering an nth-order linear scalar ordinary differential equation, it is trivially equivalent to a first-order n-dimensional linear system. In this case, the change on the boundary conditions determined by the coefficient matrices B and C does not always imply the change in the boundary conditions of the corresponding scalar equation. To see this, consider the following second-order Dirichlet boundary value problem:

$$u''(t) + m\,u(t) = f(t),\ t \in [0, 2\pi], \qquad u(0) = u(2\pi) = 0.$$

Clearly this equation corresponds to operator

$$L \begin{pmatrix} w_1(t) \\ w_2(t) \end{pmatrix} = \begin{pmatrix} w_1'(t) \\ w_2'(t) \end{pmatrix} - \begin{pmatrix} 0 & 1 \\ -m & 0 \end{pmatrix} \begin{pmatrix} w_1(t) \\ w_2(t) \end{pmatrix},$$

defined on

$$D(L) = \left\{ \begin{pmatrix} w_1 \\ w_2 \end{pmatrix} \in W, \ \begin{pmatrix} 1 & 0 \\ 0 & 0 \end{pmatrix} \begin{pmatrix} w_1(0) \\ w_2(0) \end{pmatrix} + \begin{pmatrix} 0 & 0 \\ 1 & 0 \end{pmatrix} \begin{pmatrix} w_1(2\pi) \\ w_2(2\pi) \end{pmatrix} = \begin{pmatrix} 0 \\ 0 \end{pmatrix} \right\}.$$

Obviously

$$\langle w(2\pi), z(2\pi)\rangle = \langle w(0), z(0)\rangle \text{ for all } w \in D(L) \text{ if and only if } z_2(0) = z_2(2\pi) = 0.$$

Thus

$$L^* \begin{pmatrix} z_1(t) \\ z_2(t) \end{pmatrix} = - \begin{pmatrix} z_1'(t) \\ z_2'(t) \end{pmatrix} - \begin{pmatrix} 0 & -m \\ 1 & 0 \end{pmatrix} \begin{pmatrix} z_1(t) \\ z_2(t) \end{pmatrix},$$

with

$$D(L^*) = \left\{ \begin{pmatrix} z_1 \\ z_2 \end{pmatrix} \in W, \ \begin{pmatrix} 0 & 1 \\ 0 & 0 \end{pmatrix} \begin{pmatrix} z_1(0) \\ z_2(0) \end{pmatrix} + \begin{pmatrix} 0 & 0 \\ 0 & 1 \end{pmatrix} \begin{pmatrix} z_1(2\pi) \\ z_2(2\pi) \end{pmatrix} = \begin{pmatrix} 0 \\ 0 \end{pmatrix} \right\}.$$

But, the second-order equation related to this adjoint system is exactly the same as the one we were considering at the beginning:

$$z_2''(t) + m\,z_2(t) = f(t),\ t \in [0, 2\pi], \qquad z_2(0) = z_2(2\pi) = 0.$$

On the other hand, it is enough to consider the initial and terminal value conditions to realize that, as in the vectorial case, the domains of definition of the operator L and L^* can be different.

1.4 nth-Order Differential Equations

In this section we are dealing with the study of the general two-point nth-order differential equation

$$L_n u(t) = \sigma(t), \ t \in J, \quad U_i(u) = h_i, \ i = 1, \ldots, n, \tag{1.4.1}$$

where

$$U_i(u) \equiv \sum_{j=0}^{n-1} \left(\alpha_j^i \, u^{(j)}(a) + \beta_j^i \, u^{(j)}(b) \right), \qquad i = 1, \ldots, n, \tag{1.4.2}$$

and

$$L_n u(t) \equiv u^{(n)}(t) + a_1(t) \, u^{(n-1)}(t) + \cdots + a_{n-1}(t) \, u'(t) + a_n(t) \, u(t), \quad t \in J, \tag{1.4.3}$$

being α_j^i, β_j^i, and h_i real constants for all $i = 1, \ldots, n$, and $j = 0, \ldots, n-1$, and $\sigma, a_k \in \mathcal{L}^1(J, \mathbb{R})$ for all $k = 1, \ldots, n$.

In this situation we look for solutions that belong to the space

$$W^{n,1}(J) = \{ u \in \mathcal{C}^{n-1}(J, \mathbb{R}), \ u^{(n-1)} \in \mathcal{AC}(J, \mathbb{R}) \}.$$

It is obvious that this boundary value problem is equivalent to the first-order n-dimensional system (1.2.1)–(1.2.2), with

$$x(t) = \begin{pmatrix} u(t) \\ u'(t) \\ \vdots \\ u^{(n-1)}(t) \end{pmatrix}, \quad A(t) = \left(\begin{array}{c|c} 0 & I_{n-1} \\ \hline -a_n(t) & -a_{n-1}(t) \ldots - a_1(t) \end{array} \right), \quad f(t) = \begin{pmatrix} 0 \\ \vdots \\ 0 \\ \sigma(t) \end{pmatrix},$$

$$B = \begin{pmatrix} \alpha_0^1 & \cdots & \alpha_{n-1}^1 \\ \vdots & \ddots & \vdots \\ \alpha_0^n & \cdots & \alpha_{n-1}^n \end{pmatrix}, \quad C = \begin{pmatrix} \beta_0^1 & \cdots & \beta_{n-1}^1 \\ \vdots & \ddots & \vdots \\ \beta_0^n & \cdots & \beta_{n-1}^n \end{pmatrix} \quad \text{and} \quad h = \begin{pmatrix} h_1 \\ \vdots \\ h_n \end{pmatrix}.$$

By definition, if $(u_1, \ldots, u_n)$ is a set of linearly independent solutions of $L_n u = 0$, then

$$\phi(t) = \begin{pmatrix} u_1(t) & \cdots & u_n(t) \\ u_1'(t) & \cdots & u_n'(t) \\ \vdots & \ddots & \vdots \\ u_1^{(n-1)}(t) & \cdots & u_n^{(n-1)}(t) \end{pmatrix}$$

is a fundamental matrix of this particular case of problem (1.2.1)–(1.2.2). As a consequence

$$M_\phi = B\,\phi(a) + C\,\phi(b) = \begin{pmatrix} U_1(u_1) & \cdots & U_1(u_n) \\ \vdots & \ddots & \vdots \\ U_m(u_1) & \cdots & U_m(u_n) \end{pmatrix}.$$

So, from Theorem 1.2.3, we have that problem (1.4.1)–(1.4.2) has a unique solution if and only if the determinant of the previous matrix is different from zero.

Moreover, Theorem 1.2.10 gives us the following necessary condition to ensure the uniqueness of solution of problem (1.4.1)–(1.4.2):

$$\text{rank} \begin{pmatrix} \alpha_0^1 & \cdots & \alpha_{n-1}^1 & \beta_0^1 & \cdots & \beta_{n-1}^1 \\ \vdots & \ddots & \vdots & \vdots & \ddots & \vdots \\ \alpha_0^n & \cdots & \alpha_{n-1}^n & \beta_0^n & \cdots & \beta_{n-1}^n \end{pmatrix} = n. \tag{1.4.4}$$

It is not difficult to verify that the results stated in Sect. 1.2 can be straightforward applied to the scalar case.

By assuming the uniqueness of solutions, we have, from (1.2.15), that the expression of x is obtained by means of the matrix Green's function $G : (J \times J)\backslash\{(t,t),\, t \in J\} \to \mathcal{M}_{n\times n}$. However, in this new situation, the function G gives us information about the whole set of values of x, in which the expression of u and all of its derivatives up to the $(n-1)$th-order are involved. Moreover, by considering the case $h = 0$, it is enough to take into account that the expression

$$\begin{pmatrix} u(t) \\ u'(t) \\ \vdots \\ u^{(n-1)}(t) \end{pmatrix} = \int_a^b \begin{pmatrix} G_{1,1}(t,s) & \cdots & G_{1,n}(t,s) \\ \vdots & \ddots & \vdots \\ G_{n,1}(t,s) & \cdots & G_{n,n}(t,s) \end{pmatrix} \begin{pmatrix} 0 \\ \vdots \\ 0 \\ \sigma(s) \end{pmatrix} ds,$$

tells us that

$$u(t) = \int_a^b G_{1,n}(t,s)\,\sigma(s)\,ds \equiv \int_a^b g(t,s)\,\sigma(s)\,ds. \tag{1.4.5}$$

It is important to remark that function $G_{1,n}$ is continuous at the diagonal for $n \geq 2$. Thus, to avoid tedious notation, in the sequel we will assume that this function is defined in the whole square $J \times J$, taking care of the case $n = 1$ at the diagonal.

Function $g \equiv G_{1,n} : J \times J \to \mathbb{R}$ is the so-called **Green's function** related to the nth-order linear problem (1.4.1)–(1.4.2).

Let us see now how to obtain a similar characterization to the conditions (G1)–(G6), introduced in Definition 1.2.16, for this new situation for function g.

First, note that from condition (G4) and the definition of the coefficient matrix $A(t)$, we have that for all $i \in \{1, \ldots, n\}$ and $t \neq s$, it is satisfied:

$$G_{2,i}(t, s) = \frac{\partial}{\partial t} G_{1,i}(t, s),$$

$$G_{3,i}(t, s) = \frac{\partial}{\partial t} G_{2,i}(t, s) = \frac{\partial^2}{\partial t^2} G_{1,i}(t, s),$$

$$\vdots$$

$$G_{n,i}(t, s) = \frac{\partial}{\partial t} G_{n-1,i}(t, s) = \frac{\partial^{n-1}}{\partial t^{n-1}} G_{1,i}(t, s),$$

$$\frac{\partial^n}{\partial t^n} G_{1,i}(t, s) = \frac{\partial}{\partial t} G_{n,i}(t, s) = -\sum_{j=1}^{n} a_{n-j+1}(t) G_{j,i}(t, s) = -\sum_{j=0}^{n-1} a_{n-j}(t) \frac{\partial^j}{\partial t^j} G_{1,i}(t, s).$$

So, from this property and condition (G6), we know that every element of the first row of the Green matrix function is a solution of the nth-order homogeneous problem:

$$L_n u(t) = 0 \ a.e. \ t \in J \setminus \{s\}, \quad U_i(u) = 0, \ i = 1, \ldots, n. \tag{1.4.6}$$

In particular this property holds for the scalar Green's function g.

From (G3) and (G5), and the previous properties concerning the partial derivatives of function g, we have that it is a $\mathscr{C}^{n-2}(J \times J)$ function that satisfies that

$$\lim_{s \to t^-} \frac{\partial^{n-1}}{\partial t^{n-1}} g(t, s) = \lim_{s \to t^-} \frac{\partial^{n-1}}{\partial t^{n-1}} G_{1,n}(t, s) = \lim_{s \to t^-} G_{n,n}(t, s) =$$

$$1 + \lim_{s \to t^+} G_{n,n}(t, s) = 1 + \lim_{s \to t^+} \frac{\partial^{n-1}}{\partial t^{n-1}} G_{1,n}(t, s) = 1 + \lim_{s \to t^+} \frac{\partial^{n-1}}{\partial t^{n-1}} g(t, s).$$

Thus, we are in a position to characterize a Green's function for problem (1.4.1)–(1.4.2) as follows.

Definition 1.4.1. We say that g is a *Green's function* for problem (1.4.1)–(1.4.2) if it satisfies the following properties:

(g1) g is defined on the square $J \times J$ (except $t = s$ if $n = 1$).

(g2) For $k = 0, 1, \ldots, n - 2$, the partial derivatives $\frac{\partial^k g}{\partial t^k}$ exist and they are continuous on $J \times J$.

(g3) $\frac{\partial^{n-1} g}{\partial t^{n-1}}$ and $\frac{\partial^n g}{\partial t^n}$ exist and are continuous on the triangles $a \leq s < t \leq b$ and $a \leq t < s \leq b$.

(g4) For each $s \in (a, b)$, the function $t \to g(t, s)$ is a solution of the differential equation $L_n y = 0$ a.e. on $[a, s) \cup (s, b]$. That is,

$$\frac{\partial^n}{\partial t^n} g(t, s) + a_1(t) \frac{\partial^{n-1}}{\partial t^{n-1}} g(t, s) + \cdots + a_{n-1}(t) \frac{\partial}{\partial t} g(t, s) + a_n(t) g(t, s) = 0,$$

for al $t \in J \setminus \{s\}$.

(g5) For each $t \in (a, b)$ there exist the lateral limits

$$\frac{\partial^{n-1}}{\partial t^{n-1}} g(t^-, t) = \frac{\partial^{n-1}}{\partial t^{n-1}} g(t, t^+) \quad \text{and} \quad \frac{\partial^{n-1}}{\partial t^{n-1}} g(t, t^-) = \frac{\partial^{n-1}}{\partial t^{n-1}} g(t^+, t)$$

and, moreover,

$$\frac{\partial^{n-1}}{\partial t^{n-1}} g(t^+, t) - \frac{\partial^{n-1}}{\partial t^{n-1}} g(t^-, t) = \frac{\partial^{n-1}}{\partial t^{n-1}} g(t, t^-) - \frac{\partial^{n-1}}{\partial t^{n-1}} g(t, t^+) = 1.$$

(g6) For each $s \in (a, b)$, the function $t \to g(t, s)$ satisfies the boundary conditions $U_i(g(\cdot, s)) = 0$, $i = 1, \ldots, n$, i.e.,

$$\sum_{j=0}^{n-1} \left(\alpha_j^i \frac{\partial^j}{\partial t^j} g(a, s) + \beta_j^i \frac{\partial^j}{\partial t^j} g(b, s) \right) = 0, \qquad i = 1, \ldots, n.$$

As in Theorem 1.2.17 and Corollary 1.2.4, it is immediate to prove the following result.

Theorem 1.4.2. *Problem (1.4.1) has a unique solution if and only if there exists a unique Green's function related to this problem.*

In Examples 1.2.18–1.2.20 it is shown that if the system (1.2.3) has nontrivial solutions, then the uniqueness of Green's function of the related system cannot be ensured. It is immediate to adapt such examples to the second-order equation $u''(t) + u(t) = \sigma(t)$, $t \in [0, 2\pi]$, coupled with either the periodic conditions $u(0) = u(2\pi)$, $u'(0) = u'(2\pi)$ (Example 1.2.18), or $u(0) + u(2\pi) = 0$ (Example 1.2.19), or the Neumann ones $u'(0) = u'(2\pi) = 0$ (Example 1.2.20).

Moreover, Lemma 1.2.21 has an immediate adaptation to this situation.

Arguing as in the vectorial case, one can verify that when considering the initial value problem ($\alpha_j^i = \delta_{i, i-1}$, $\beta_j^i = 0$, $i = 1, \ldots, n$, $j = 0, \ldots, n-1$, with $\delta_{l,k}$ the Kronecker delta function), then $g(\cdot, a) : J \to \mathbb{R}$ is the unique solution of the problem

$$L_n r(t) = 0, \ t \in J, \quad r^{(i)}(a) = 0, \ i = 0, \ldots, n-2, \quad r^{(n-1)}(a) = 1. \quad (1.4.7)$$

If we study the terminal value problem ($\alpha_j^i = 0$, $\beta_j^i = \delta_{i, i-1}$, $i = 1, \ldots, n$, $j = 0, \ldots, n-1$), then $g(\cdot, b) : J \to \mathbb{R}$ is the unique solution of the problem

$$L_n r(t) = 0, \ t \in J, \quad r^{(i)}(b) = 0, \ i = 0, \ldots, n-2, \quad r^{(n-1)}(b) = -1. \quad (1.4.8)$$

Moreover, if the periodic problem is considered ($\alpha_j^i = \delta_{i,i-1}$, $\beta_j^i = -\delta_{i,i-1}$, $i = 1,\ldots,n$, $j = 0,\ldots,n-1$), we have that $g(\cdot,a) : J \to \mathbb{R}$ is the unique solution of the problem

$$L_n\, r(t)=0, \ t \in J, \ r^{(i)}(a)=r^{(i)}(b), \ i = 0,\ldots,n-2, r^{(n-1)}(a)=r^{(n-1)}(b) + 1.$$

$$(1.4.9)$$

If the coefficients a_j, $j = 1,\ldots,n$, involved in the definition of the operator L_n are constant, arguing as in Sect. 1.2, we can deduce sharper properties of the studied Green's functions. For instance, if we consider the initial value problem, then Green's function satisfies

$$g(t,s) = g(t-s+a,a), \quad \text{if } a \leq s \leq t \leq b \text{ and } g(t,s) = 0 \text{ otherwise.} \quad (1.4.10)$$

Concerning the terminal value problem we have that

$$g(t,s) = g(b+t-s,b), \quad \text{if } a \leq t \leq s \leq b \text{ and } g(t,s) = 0 \text{ otherwise.} \quad (1.4.11)$$

Finally, if we consider the periodic case, we deduce that

$$g(t,s)=g(a+t-s,a), \quad \text{if } a \leq s \leq t \leq b, \text{ and } g(t,s)=g(b+t-s,a), \text{ otherwise.}$$

$$(1.4.12)$$

As in the matrix case, if the coefficients of operator L_n are constants, the functions related to these three problems are constant over the straight lines of slope one. This property allows us to obtain the expression of Green's function by avoiding tedious calculations. Moreover, it gives us the opportunity of attaining fundamental results related with its sign or suitable a priori bounds.

Example 1.4.3. Consider, for any $m \in \mathbb{R}$, the second-order initial value problem

$$u''(t) + m\,u(t) = \sigma(t), \quad u(0) = u'(0) = 0.$$

To obtain Green's function we only need to solve the following problem:

$$r''(t) + m\,r(t) = 0, \ t \in \mathbb{R}, \quad r(0) = 0, \ r'(0) = 1.$$

It is immediate to verify that

$$r(t) = \begin{cases} \dfrac{\sin(\sqrt{m}\,t)}{\sqrt{m}}, & \text{if } m > 0, \\ t, & \text{if } m = 0, \\ \dfrac{\sinh(\sqrt{-m}\,t)}{\sqrt{-m}}, & \text{if } m < 0. \end{cases}$$

So, since $r(t) = g(t,0)$, the expression of Green's function is deduced from expression (1.4.10).

If we are interested in the periodic case,

$$u''(t) + m\, u(t) = \sigma(t), \quad u(0) = u(1),\ u'(0) = u'(1).$$

The expression of Green's function is obtained by solving problem (1.4.9), i.e.,

$$r''(t) + m\, r(t) = 0,\ t \in \mathbb{R}, \quad r(0) = r(1),\ r'(0) = r'(1) + 1.$$

Such equation has a unique solution given by

$$r(t) = \begin{cases} \dfrac{\cos(\sqrt{m}\,(t-\frac{1}{2}))}{2\sqrt{m}\,\sin\left(\frac{\sqrt{m}}{2}\right)}, & \text{if } m \neq 4k^2\,\pi^2,\ k = 0,1,\ldots \\[2ex] -\dfrac{\cosh(\sqrt{-m}\,(t-\frac{1}{2}))}{2\sqrt{-m}\,\sinh\left(\frac{\sqrt{-m}}{2}\right)}, & \text{if } m < 0. \end{cases}$$

We remark that the expression of function r tells us the exact values of the parameter $m \in \mathbb{R}$ for which the periodic problem is not uniquely solvable. In such a case, we have not a unique solution if and only if $m = 4k^2\,\pi^2,\ k = 0,1,\ldots$ Such values are known as the eigenvalues of the operator u'' on the space of the 1-periodic functions.

In general, even in the case of constant coefficients a_j, Green's function is not constant along the straight line of slope which equals to one, so we cannot deduce the expression of Green's function by means of its values at $s = a$ or $s = b$. In that case, to obtain the expression of the function, we need to use the conditions (g1)–(g6) that characterize these functions in Definition 1.4.1. In the next example we study the Neumann case for a second-order equation. Similar arguments may be found in [42].

Example 1.4.4. Given $m > 0$, $T > 0$ and $\sigma \in \mathscr{L}^1([0, T], \mathbb{R})$, we are interested in constructing Green's function of the following second-order problem with Neumann boundary value conditions:

$$u''(t) + m^2\, u(t) = \sigma(t),\ t \in [0, T], \quad u'(0) = u'(T) = 0.$$

First, having in mind conditions (g1)–(g6) introduced in Definition 1.4.1, we look for a two-variable function $g(t, s)$, continuous on $[0, T] \times [0, T]$, such that $\partial g/\partial t$ and $\partial^2 g/\partial t^2$ are continuous in the triangles $0 \leq s < t \leq T$ and $0 \leq t < s \leq T$ and that satisfies the equation

$$\frac{\partial^2}{\partial t^2} g(t, s) + m^2\, g(t, s) = 0, \quad \text{for all } t \neq s.$$

So, we have that

$$g(t, s) = \begin{cases} A(s) \sin m\, t + B(s) \cos m\, t, & \text{if } 0 \leq s < t \leq T, \\ C(s) \sin m\, t + D(s) \cos m\, t, & \text{if } 0 \leq t < s \leq T. \end{cases}$$

Since function g is continuous on (t, t), we deduce

$$A(t) \sin m\,t + B(t) \cos m\,t = C(t) \sin m\,t + D(t) \cos m\,t.$$

Now, from condition (g5)

$$\lim_{s \to t^-} \frac{\partial}{\partial t} g(t, s) - \lim_{s \to t^+} \frac{\partial}{\partial t} g(t, s) = 1,$$

we have

$$m\,A(t) \cos m\,t - m\,B(t) \sin m\,t - m\,C(t) \cos m\,t + m\,D(t) \sin m\,t = 1.$$

The expressions of the coefficient functions are obtained from condition (g6) as follows:

Equality $\frac{\partial}{\partial t} g(0, s) = 0$ means $C(t) = 0$.

Moreover, $\frac{\partial}{\partial t} g(T, s) = 0$ is equivalent to

$$A(t) \cos m\,T = B(t) \sin m\,T.$$

So we have constructed an algebraic linear system of four equations with four variables. It is immediate to deduce that

$$A(t) = \frac{\cos m\,t}{m}, \quad B(t) = \frac{\cos m\,T \cos m\,t}{m \sin m\,T} \quad \text{and} \quad D(t) = \frac{\cos m\,(T - t)}{m \sin m\,T}.$$

As a consequence, we conclude that the related Green's function has the expression

$$g(t, s) = \frac{1}{m \sin m\,T} \begin{cases} \cos m\,s \cos m\,(T - t), & \text{if } 0 \le s < t \le T, \\ \cos m\,t \cos m\,(T - s), & \text{if } 0 \le t < s \le T. \end{cases}$$

As in the previous example of the periodic case, we have that this problem has not a unique solution if and only if $\sin m\,T = 0$, i.e., $m = k\,\pi/T$, $k = 1, 2 \ldots$

In a similar way, one can prove that this problem has not a unique solution when $m = 0$. However, if we consider the case

$$u''(t) - m^2\,u(t) = \sigma(t), \ t \in [0, T], \quad u'(0) = u'(T) = 0,$$

we can deduce the uniqueness of solutions for all $m > 0$ and the corresponding expression of Green's function by analogous arguments.

Remark 1.4.5. For a general nth-order differential equation the calculations that are needed to obtain the expression of Green's function are very complicated. On Appendix A of this book it is shown a Mathematica code that has been developed in [19] to deduce the expression of Green's function related to a two-point boundary value problem.

As in the vectorial case, we can consider multipoint boundary conditions of the form

$$\sum_{j=0}^{n-1}\sum_{k=0}^{m}\alpha_{j,k}^{i}\,u^{(j)}(c_k) = h_i, \qquad i = 1,\ldots,n,$$

with m a positive integer, $\alpha_{j,k}^{i}$ and h_i real constants for all $i = 1,\ldots,n$, $k = 0,\ldots,m$ and $j = 0,\ldots,n-1$ and $a = c_0 < c_1 < \ldots < c_m = b$.

The existence and uniqueness of solutions in this new situation follow by considering the associated first-order n-dimensional system and arguing as in Lemma 1.2.23.

On the other hand, in the scalar case, in order to deal with the adjoint of an operator defined on a Hilbert space, we consider the space $\mathscr{L}^2(J,\mathbb{R})$, which is a Hilbert space with the inner product

$$(u,v)_2 := \int_a^b u(t)\,v(t)\,dt.$$

Now, by considering a two-variable real function $k : J \times J \to \mathbb{R}$, such that $k \in \mathscr{L}^2(J \times J,\mathbb{R})$, we define the integral operator $T : \mathscr{L}^2(J,\mathbb{R}) \to \mathscr{L}^2(J,\mathbb{R})$ as follows:

$$T\,u(t) = \int_a^b k(t,s)\,u(s)\,ds, \qquad t \in J. \tag{1.4.13}$$

Arguing as in Theorem 1.3.1, we can conclude that its adjoint operator T^* is given by the following expression:

$$T^*\,v(t) = \int_a^b k(s,t)\,v(s)\,ds, \qquad t \in J. \tag{1.4.14}$$

In particular, the operator T is self-adjoint if and only if $k(t,s) = k(s,t)$ for a.e. $(t,s) \in J \times J$.

In order to conserve a Hilbert space structure, in this situation, the solutions of differential equations belong to the space

$$W^{n,2}(J) = \{u \in \mathscr{C}^{n-1}(J,\mathbb{R}),\ u^{(n-1)} \in \mathscr{AC}(J,\mathbb{R}),\ u^{(n)} \in \mathscr{L}^2(J,\mathbb{R})\}.$$

When the integral operator T is the inverse of the differential one L_n defined in a suitable domain $D(L_n)$, to obtain the expression of the nth-order differential operator L_n^*, which is the inverse of the adjoint operator T^*, we will assume that the coefficients a_k belong to $\mathscr{C}^{n-k}(J)$ and take into account the following equality, deduced immediately from integration by parts:

$$\int_a^b a_{n-j}(t)\, u^{(j)}(t)\, v(t)\, dt = (-1)^j \int_a^b (a_{n-j}\, v)^{(j)}(t)\, u(t)\, dt$$

$$+ \sum_{i=0}^{j-1} (-1)^{j-1-i}\, (a_{n-j}\, v)^{(j-1-i)}(b)\, u^{(i)}(b)$$

$$- \sum_{i=0}^{j-1} (-1)^{j-1-i}\, (a_{n-j}\, v)^{(j-1-i)}(a)\, u^{(i)}(a).$$

Here $j \in \{1,\ldots,n\}$, $v \in \mathscr{C}^k(J,\mathbb{R})$ and we set $a_0(t) \equiv 1$.

As a consequence, from the definition of the adjoint operator, and arguing as in Sect. 1.3, we conclude that the adjoint differential operator is given by the expression

$$L_n^* v(t) = (-1)^n v^{(n)}(t) + \sum_{j=1}^{n-1} (-1)^j\, (a_{n-j}\, v)^{(j)}(t) + a_n(t)\, v(t), \quad \text{for all } v \in D(L_n^*),$$

(1.4.15)

and

$$D(L_n^*) = \left\{ v \in W^{n,2}(J,\mathbb{R}),\ \text{such that} \sum_{j=1}^{n}\sum_{i=0}^{j-1} (-1)^{j-1-i}\, (a_{n-j}\, v)^{(j-1-i)}(b)\, u^{(i)}(b) \right.$$

$$\left. = \sum_{j=1}^{n}\sum_{i=0}^{j-1} (-1)^{j-1-i}\, (a_{n-j}\, v)^{(j-1-i)}(a)\, u^{(i)}(a)\ (\text{with } a_0 = 1), \quad \text{for all } u \in D(L_n) \right\}.$$

(1.4.16)

Example 1.4.6. Consider the second-order operator

$$L\, u(t) = u''(t) + t\, u'(t) + (\sin t)\, u(t),$$

defined on the space

$$D(L) = \{ u \in W^{2,2}([0,1],\mathbb{R}),\quad u'(0) = u'(1) = 0 \}.$$

In this case we have that

$$L^* v(t) = v''(t) - (t\, v(t))' + (\sin t)\, v(t) = v''(t) - t\, v'(t) + (-1 + \sin t)\, v(t).$$

The set of definition $D(L^*)$ of the adjoint operator consists of the functions $v \in W^{2,2}([0,1],\mathbb{R})$ that satisfy the following equality for all $u \in D(L)$:

$$(t\, v(t) - v'(t))\, u(t) + v(t)\, u'(t)|_{t=0} = (t\, v(t) - v'(t))\, u(t) + v(t)\, u'(t)|_{t=1}.$$

Due to the fact that $u \in D(L)$ implies $u'(0) = u'(1)$, we conclude that the previous equality holds if and only if

$$-v'(0) u(0) = (v(1) - v'(1)) u(1) \qquad \text{for all } u \in D(L).$$

That is

$$D(L^*) = \{v \in W^{2,2}([0, 1], \mathbb{R}), v'(0) = v(1) - v'(1) = 0\}.$$

Notice that if, instead of the Neumann boundary conditions, we study the Dirichlet case

$$D(L) = \{u \in W^{2,2}([0, 1], \mathbb{R}), u(0) = u(1) = 0\}$$

then

$$D(L^*) = \{v \in W^{2,2}([0, 1], \mathbb{R}), v(0) = v(1) = 0\}.$$

For the periodic case

$$D(L) = \{u \in W^{2,2}([0, 1], \mathbb{R}), u(0) = u(1); \ u'(0) = u'(1)\},$$

we conclude

$$D(L^*) = \{v \in W^{2,2}([0, 1], \mathbb{R}), v(0) = v(1) = v'(1) - v'(0)\}.$$

Remark 1.4.7. In Sect. 1.3 we have shown that the nth-order differential operator L_n^* corresponds with the first-order n-dimensional one:

$$L^*z(\cdot) := -z'(\cdot) - A^T(\cdot)\, z(\cdot) = \begin{pmatrix} -\frac{d}{dt} & 0 & 0 & \cdots & \cdots & 0 & a_n(\cdot) \\ -1 & -\frac{d}{dt} & 0 & \cdots & \cdots & 0 & a_{n-1}(\cdot) \\ 0 & -1 & -\frac{d}{dt} & \ddots & \ddots & 0 & a_{n-2}(\cdot) \\ \vdots & \vdots & \ddots & \ddots & \ddots & \vdots & \vdots \\ 0 & \cdots & 0 & -1 & -\frac{d}{dt} & 0 & a_3(\cdot) \\ 0 & \cdots & \cdots & 0 & -1 & -\frac{d}{dt} & a_2(\cdot) \\ 0 & \cdots & \cdots & \cdots & 0 & -1 & -\frac{d}{dt} + a_1(\cdot) \end{pmatrix} z(\cdot).$$

To construct the nth-order differential equation that defines the previous operator, we only need to calculate the determinant d_n of the previous matrix. It is obvious that it is given by the following recurrence expression:

$$d_n = -\frac{d}{dt} d_{n-1} + a_n, \ n \geq 2, \qquad d_1 = -\frac{d}{dt} + a_1.$$

It is easy to verify that the unique solution of this difference equation is given by the following expression:

$$d_n = (-1)^n \frac{d^n}{d\,t^n} + \sum_{j=1}^{n-1} (-1)^j \frac{d^j}{d\,t^j} a_{n-j} + a_n,$$

which has a corresponding operator $d_n(z)$ that coincides with the definition of the differential operator L_n^*.

It is important to point out that when the coefficients a_k are constants for all $k \in \{1, \ldots, n-1\}$, then we have that

$$L_n^* v(t) = (-1)^n v^{(n)}(t) + \sum_{j=1}^{n-1} (-1)^j a_{n-j}\, v^{(j)}(t) + a_n(t)\, v(t), \quad \text{for all } v \in D(L_n^*).$$

Moreover, we deduce the following necessary condition to ensure the self-adjoint character of a differential operator.

Corollary 1.4.8. *If the operator L_n is self-adjoint, and its coefficients are constant, then n must be even and $a_k = 0$ for all k odd.*

Example 1.4.9. It is not difficult to verify that the second-order operator $u''(\cdot) + a_2(\cdot)\, u(\cdot)$ is self-adjoint when the so-called separated Sturm-Liouville boundary conditions $p_0\, u(a) - q_0\, u'(a) = p_1\, u(b) + q_1\, u'(b) = 0$, with p_0, p_1, q_0, $q_1 \geq 0$, $p_0 + q_0 > 0$ and $p_1 + q_1 > 0$, are considered. We note that $p_0 = q_0 = 0$ give us the Neumann boundary conditions and that $p_1 = q_1 = 0$ are the Dirichlet ones.

The same holds if we study the periodic boundary value conditions $u(a) = u(b)$, $u'(a) = u'(b)$. This property remains valid for the operator $u^{(2n)}(\cdot) + a_{2n}(\cdot)\, u(\cdot)$ with periodic conditions $u^{(i)}(a) = u^{(i)}(b)$, $i = 0, \ldots, n-1$.

We remark that the same operator is not self-adjoint if the boundary conditions are either the initial $u(a) = u'(a) = 0$ or the terminal $u(b) = u'(b) = 0$.

When we are dealing with the periodic boundary conditions, we have the following property.

Proposition 1.4.10. *Suppose that the periodic boundary value problem*

$$L_n\, u(t) = \sigma(t), \quad t \in J, \quad u^{(i)}(a) = u^{(i)}(b), \ i = 0, \ldots, n-1,$$

has a unique solution for all $\sigma \in \mathscr{L}^1(J, \mathbb{R})$. By denoting g as its related Green's function, we have that

$$\int_a^b g(t, s)\, a_n(s)\, ds = 1.$$

Proof. First we note that if $a_n(t) = 0$ for all $t \in J$, then every constant function is a solution of the homogeneous equation $L_n u = 0$ and, as a consequence, the considered problem has not a unique solution.

The proof is a direct consequence of the definition of a Green's function and the fact that $u(t) = 1$ for all $t \in J$ is the unique solution of problem

$$L_n u(t) = a_n(t), \quad t \in J, \quad u^{(i)}(a) = u^{(i)}(b), \ i = 0, \dots, n-1. \qquad \square$$

We note that the same property holds for any problem in which the constant 1 is the unique solution of the equation $L_n u(t) = a_n(t)$. This is the case, for instance, of the second-order Neumann problem

$$u''(t) + a_1 u'(t) + a_2(t) u(t) = \sigma(t), \quad t \in J, \quad u'(a) = u'(b) = 0.$$

For the periodic case, when the coefficients involved in operator L_n are constant, we are able to deduce some additional properties of symmetry.

Proposition 1.4.11. *Suppose that the periodic boundary value problem*

$$\begin{cases} u^{(2n)}(t) + a_2 u^{(2n-2)}(t) + \cdots + a_{2n-2} u''(t) + a_{2n} u(t) = \sigma(t), \quad t \in J, \\ u^{(i)}(a) = u^{(i)}(b), \ i = 0, \dots, 2n-1, \end{cases}$$

has a unique solution for all $\sigma \in \mathcal{L}^1(J, \mathbb{R})$. Then the related Green's function g satisfies the following symmetry condition:

$$g(t, s) = g(a + b - t, a + b - s) \quad \text{for all } t, s \in J.$$

Proof. As we have previously shown, $r(\cdot) := g(\cdot, a)$ is the unique solution of the problem

$$\begin{cases} r^{(2n)}(t) + a_2 r^{(2n-2)}(t) + \cdots + a_{2n-2} r''(t) + a_{2n} r(t) = 0, \quad t \in J, \\ r^{(i)}(a) = r^{(i)}(b), \ i = 1, \dots, 2n-2, \quad r^{(2n-1)}(a) = r^{(2n-1)}(b) + 1. \end{cases}$$

It is immediate to verify that $r(t) = r(a + b - t)$ for all $t \in J$.

As a consequence, suppose that $a \le t \le s \le b$, then from (1.4.12) we have that

$$g(a + b - t, a + b - s) = g(a + s - t, a) = g(s, t).$$

The proof follows from the fact that this operator is self-adjoint, which is equivalent to the fact that $g(s, t) = g(t, s)$.

The case $a \le s \le t \le b$ is analogous. $\qquad \square$

As a direct consequence we attain at the following result.

Corollary 1.4.12. *If the periodic boundary value problem*

$$\begin{cases} u^{(2n)}(t) + a_2 \, u^{(2n-2)}(t) + \cdots + a_{2n-2} \, u''(t) + a_{2n} \, u(t) = \sigma(t), \quad t \in J, \\ u^{(i)}(a) = u^{(i)}(b), \; i = 0, \ldots, 2n-1, \end{cases}$$

has a unique solution for all $\sigma \in \mathscr{L}^1(J, \mathbb{R})$, then the following properties hold for the related Green's function g:

1. For all $j = 0, \ldots, n-2$, it is fulfilled

$$\frac{\partial^{2j+1}}{\partial t^{2j+1}} g(t^+, t) = \frac{\partial^{2j+1}}{\partial t^{2j+1}} g(t, t^-) = \frac{\partial^{2j+1}}{\partial t^{2j+1}} g(t, t^+) = \frac{\partial^{2j+1}}{\partial t^{2j+1}} g(t^-, t) = 0.$$

2.

$$\frac{\partial^{2n-1}}{\partial t^{2n-1}} g(t^+, t) = \frac{\partial^{2n-1}}{\partial t^{2n-1}} g(t, t^-) = \frac{1}{2}, \quad \frac{\partial^{2n-1}}{\partial t^{2n-1}} g(t^-, t) = \frac{\partial^{2n-1}}{\partial t^{2n-1}} g(t, t^+) = -\frac{1}{2}.$$

Proof. The definition of function $r(\cdot) := g(\cdot, a)$, together with the symmetric condition of function proved in Proposition 1.4.11, tells us that for all odd number $i \leq 2n-3$, it is satisfied that $r^{(i)}(a) = r^{(i)}(b) = -r^{(i)}(a)$, which implies that $r^{(i)}(a) = r^{(i)}(b) = 0$.

On the other hand, $-r^{(2n-1)}(b) = r^{(2n-1)}(a) = r^{(2n-1)}(b) + 1$, so we conclude that $r^{(2n-1)}(a) = 1/2$ and $r^{(2n-1)}(b) = -1/2$.

From (1.4.12) we have that

$$\frac{\partial^j}{\partial t^j} g(t^+, t) = \frac{\partial^j}{\partial t^j} g(t, t^-) = r^{(j)}(a), \quad \text{for all } j = 0, \ldots, 2n-1$$

and

$$\frac{\partial^j}{\partial t^j} g(t^-, t) = \frac{\partial^j}{\partial t^j} g(t, t^+) = r^{(j)}(b), \quad \text{for all } j = 0, \ldots, 2n-1,$$

and the assertion holds.   □

If the order of the equation is odd, the operator L_n is not self-adjoint. Despite this, we can deduce some relationship between Green's functions of different pairs of operators.

Proposition 1.4.13. *Suppose that the periodic boundary value problem*

$$\begin{cases} u^{(2n+1)}(t) + \displaystyle\sum_{j=1}^{n} a_{2j} \, u^{(2(n-j)+1)}(t) + a_{2n+1} \, u(t) = \sigma(t), \quad t \in J, \\ u^{(i)}(a) = u^{(i)}(b), \; i = 0, \ldots, 2n, \end{cases}$$

has a unique solution for all $\sigma \in \mathcal{L}^1(J, \mathbb{R})$. Then the same holds for the problem

$$
\begin{cases}
u^{(2n+1)}(t) + \displaystyle\sum_{j=1}^{n} a_{2j}\, u^{(2(n-j)+1)}(t) - a_{2n+1}\, u(t) = \sigma(t), \quad t \in J, \\[4mm]
u^{(i)}(a) = u^{(i)}(b), \quad i = 0, \ldots, 2n.
\end{cases}
$$

Moreover, the related Green's functions g_1 and g_2 satisfy the following symmetry condition:

$$
g_1(t, s) = -g_2(a + b - t, a + b - s) \quad \text{for all } t, s \in J.
$$

Proof. In this case, $r(\cdot) := g_1(\cdot, a)$ is the unique solution of the problem

$$
\begin{cases}
r^{(2n+1)}(t) + a_2\, r^{(2n-1)}(t) + \cdots + a_{2n}\, r'(t) + a_{2n+1}\, r(t) = 0, \quad t \in J, \\
r^{(i)}(a) = r^{(i)}(b), \ i = 1, \ldots, 2n - 1, \quad r^{(2n)}(a) = r^{(2n)}(b) + 1.
\end{cases}
$$

It is immediate to verify that $g_2(t, a) = -r(a + b - t)$ for all $t \in J$.
As a consequence, suppose that $a \le t \le s \le b$, then, by (1.4.12),

$$
\begin{aligned}
g_2(a + b - t, a + b - s) &= \ g_2(a + s - t, a) = -r(b + t - s) \\
&= -g_1(b + t - s, a) = -g_1(t, s).
\end{aligned}
$$

The case $a \le t \le s \le b$ holds in a similar way. $\square$

We finish this section with the following property for the periodic problem with constant coefficients. This general result improves the one shown in [17, Proposition 3.1] for the second-order case $u'' + m^2 u$.

Proposition 1.4.14. *Assume $n \ge 2$ and that $a_k \in \mathbb{R}$ for all $k \in \{1, \ldots, n\}$. Suppose that the periodic boundary value problem*

$$
L_n u(t) = \sigma(t), \quad t \in J, \quad u^{(i)}(a) = u^{(i)}(b), \ i = 0, \ldots, n - 1,
$$

has a unique solution for all $\sigma \in \mathcal{L}^1(J, \mathbb{R})$. Then the following property is fulfilled:

$$
\frac{\partial g}{\partial t}(t, s) = -\frac{\partial g}{\partial s}(t, s), \quad \text{for all } t, s \in J.
$$

Proof. Suppose that function σ is differentiable. Let u be a solution of the considered periodic problem, then $u \in \mathscr{C}^n(J, \mathbb{R})$ and $v \equiv u'$ is a solution of

$$
\begin{aligned}
L_n v(t) &= \sigma'(t), & t \in J, \\
v^{(i)}(a) - v^{(i)}(b) &= 0, & i = 0, \ldots, n - 2 \\
v^{(n-1)}(a) - v^{(n-1)}(b) &= \sigma(a) - \sigma(b).
\end{aligned}
$$

Therefore, from the properties of Green's function of the periodic problem shown in this section, we deduce that

$$v(t) = \int_a^b g(t,s)\,\sigma'(s)\,ds + g(t,a)\,(\sigma(a) - \sigma(b)).$$

So, by integration by parts and from the fact that $g(t,a) = g(t,b)$, we have that

$$v(t) = g(t,b)\,\sigma(b) - g(t,a)\,\sigma(a) - \int_a^t \frac{\partial g}{\partial s}(t,s)\,\sigma(s)\,ds - \int_t^b \frac{\partial g}{\partial s}(t,s)\,\sigma(s)\,ds$$

$$+ g(t,a)\,(\sigma(a) - \sigma(b)) = -\int_a^b \frac{\partial g}{\partial s}(t,s)\,\sigma(s)\,ds.$$

On the other hand, using that $n \geq 2$, we deduce

$$v(t) = u'(t) = \frac{\partial}{\partial t}\int_a^t g(t,s)\,\sigma(s)\,ds + \frac{\partial}{\partial t}\int_t^b g(t,s)\,\sigma(s)\,ds = \int_a^b \frac{\partial g}{\partial t}(t,s)\,\sigma(s)\,ds.$$

Since the differentiable functions are dense in $\mathscr{L}^2(J,\mathbb{R})$, we conclude that

$$\frac{\partial G}{\partial t}(t,s) = -\frac{\partial G}{\partial s}(t,s). \qquad \square$$

Sometimes the difficulty of the calculations to be made in the study of Green's function depends strongly on the extremes of the interval. In general it is easier to obtain the explicit expression of the considered problem in, for instance, the intervals $[0, 1]$ or $[0, 2\pi]$ than in the general one $[a,b]$. In the next result we show that if the linear operator L_n has constant coefficients, by means of a simple change of variables, we can choose the interval where we can deal with. The arguments extend to the general situation the case studied in [6, Lemma 2.4]. To this end we define

$$\tilde{L}_n\,u(t) \equiv u^{(n)}(t) + \tilde{a}_1\,u^{(n-1)}(t) + \cdots + \tilde{a}_{n-1}\,u'(t) + \tilde{a}_n\,u(t), \quad t \in [c,d].$$

Here

$$\tilde{a}_j = a_j\left(\frac{b-a}{d-c}\right)^j, \qquad j = 1,\ldots,n,$$

and

$$\tilde{U}_i(u) \equiv \sum_{j=0}^{n-1}\left(\tilde{\alpha}_j^i\,u^{(j)}(c) + \tilde{\beta}_j^i\,u^{(j)}(d)\right), \qquad i = 1,\ldots,n,$$

with

$$\tilde{\alpha}_j^i = \alpha_j^i\left(\frac{b-a}{d-c}\right)^{n-j}, \quad \tilde{\beta}_j^i = \beta_j^i\left(\frac{b-a}{d-c}\right)^{n-j}, \quad j = 1,\ldots,n-1, \quad i = 1,\ldots,n.$$

Lemma 1.4.15. *Suppose that the general nth-order linear differential operator L_n defined in (1.4.3) has constant coefficients. Then problem (1.4.1) has a unique solution for every $\sigma \in \mathcal{L}^1([a, b], \mathbb{R})$ and $h_i \in \mathbb{R}$, $i = 1, \ldots, n$ if and only if problem*

$$\tilde{L}_n u(t) = \tilde{\sigma}(t), \ t \in [c, d], \quad \tilde{U}_i(u) = h_i, \ i = 1, \ldots, n,$$

has a unique solution for every $\tilde{\sigma} \in \mathcal{L}^1([c, d], \mathbb{R})$.

Moreover, denoting as g and $\tilde{g}$ the corresponding related Green's functions, we have that the following equality is fulfilled:

$$\tilde{g}(t, s) = \left(\frac{d-c}{b-a}\right)^{n-1} g\left(\frac{b-a}{d-c}(t-c) + a, \frac{b-a}{d-c}(s-c) + a\right) \text{ for all } (t, s) \in [c, d] \times [c, d].$$

Proof. First note that $\tilde{\sigma} \in \mathcal{L}^1([c, d], \mathbb{R})$ if and only if there is $\sigma \in \mathcal{L}^1([a, b], \mathbb{R})$ such that

$$\tilde{\sigma}(t) = \sigma\left(\frac{b-a}{d-c}(t-c) + a\right) \quad \text{for all } t \in [c, d].$$

The first part of the proof follows from this fact and the direct verification that u is a solution of (1.4.1) if and only if

$$v(t) = \left(\frac{d-c}{b-a}\right)^n u\left(\frac{b-a}{d-c}(t-c) + a\right), \qquad t \in [c, d],$$

satisfies that

$$\tilde{L}_n v(t) = \tilde{\sigma}(t), \ t \in [c, d], \quad \tilde{U}_i(v) = h_i, \ i = 1, \ldots, n.$$

The second part of the proof is given by the following equalities for all $t \in [c, d]$:

$$v(t) = \left(\frac{d-c}{b-a}\right)^n u\left(\frac{b-a}{d-c}(t-c) + a\right)$$

$$= \left(\frac{d-c}{b-a}\right)^n \int_a^b g\left(\frac{b-a}{d-c}(t-c) + a, s\right) \sigma(s)\, ds$$

$$= \left(\frac{d-c}{b-a}\right)^{n-1} \int_c^d g\left(\frac{b-a}{d-c}(t-c) + a, \frac{b-a}{d-c}(\tau - c) + a\right) \sigma\left(\frac{b-a}{d-c}(\tau - c) + a\right) d\tau.$$

$\square$

1.5 Lower and Upper Solutions

The method of lower and upper solutions is a classical tool in the theory of nonlinear boundary value problems. It allows us to ensure the existence of a solution of the considered problem lying between a pair of ordered functions that satisfy some suitable inequalities. In particular, we have information not only about the existence of solutions but also about the location of some of them. Unfortunately, there is no direct way of constructing the pair of solutions we are looking for. Anyway, there are a lot of papers on the application of this method to different problems. Moreover, weaker regularity assumptions on their definition have been investigated in order to enlarge the set of suitable lower and upper solutions and, as a consequence, to make them easier to construct. Although this tool has been developed to ensure the existence of solutions of nonlinear problems, we introduce this method here to present different examples that point out the deep influence that the existence and uniqueness of Green's function of a related linear operator has on the existence of solution of nonlinear boundary value problems. As we will point out in some particular cases, the existence results for nonlinear problems follow when the related Green's function of a suitable linear operator has constant sign.

Next some historical notes concerning this method, and compiled on the survey [10], are shown.

The theory of lower and upper solutions was started in a paper of Picard of 1890 [45] for Partial Differential Equations and, in another one of 1893, in [46] for Ordinary Differential Equations. In both cases the existence of a solution is guaranteed using a monotone iterative technique. Existence of solutions for Cauchy equations were proved by Perron in 1915 [44]. In 1926 Müller extended Perron's results to initial value systems in [40].

Scorza Dragoni [51] introduces in 1931 the notion of lower and upper solutions for Ordinary Differential Equations with Dirichlet boundary value conditions. In particular, the author proves the existence of a solution of the considered problem lying between a lower solution α and an upper solution β such that $\alpha \leq \beta$ (well ordered).

Following the ideas of these first papers, this method has been applied in a huge number of works to different first-, second-, and higher order Ordinary Differential Equations with different type of boundary conditions. Also, Partial Differential Equations, of first and second order, have been treated in the literature.

In the classical books of Bernfeld and Lakshmikantham [3] and Ladde et al. [32] the classical theory of the method of lower and upper solutions and the monotone iterative technique are developed. The iterative technique ensures the existence of the greatest and the smallest solutions lying between the lower and the upper solutions. Moreover, both solutions are obtained as the limit of two monotone sequences that start at the lower and the upper solutions and of which the recurrence formula consists of solutions of related linear problems. We refer to the reader to the classical works of Mawhin [34–37] and Fabry and Habets [26], and the surveys in this field of De Coster and Habets [22, 23] and their monograph [24], where one can find historical and bibliographical references together with recent results and open problems.

To give an idea of the applicability of this method and its relationship with the theory of Green's functions, we present a simple problem that will help us to point out the arguments used in this theory. It is important to mention that the bigger the generality of the problem the bigger the difficulty of the arguments involved.

So, let us consider the first-order nonlinear periodic boundary value problem

$$u'(t) = f(t, u(t)), \ t \in J, \quad u(a) = u(b), \tag{1.5.1}$$

with $f : J \times \mathbb{R} \to \mathbb{R}$ a continuous function.

As a consequence, we are looking for solutions $u \in \mathscr{C}^1(J, \mathbb{R})$.

As we have remarked in the previous sections, we must take into account that on the contrary to the initial value case, in which the Cauchy theorem [28] ensures the existence of at least one solution, that is defined in an interval containing the initial point in its interior, when we impose conditions on the solutions at the boundary of the interval of definition, the existence of such solutions is not guaranteed. In fact, the concept of "local solution" has no sense. This is the case, for instance, of problem (1.5.1) with $f(t, x) = t$. Indeed, if there is a solution of this problem, then

$$u(t) = u(a) + \frac{t^2 - a^2}{2}, \quad \text{for all } t \in J.$$

Therefore

$$u(b) - u(a) = \frac{b^2 - a^2}{2} \neq 0,$$

and we deduce that this particular case of problem (1.5.1) has no solution.

Moreover, we cannot expect, in case of existence of solutions, that, in general, there is only one. Consider, for instance, the function $f(t, x) = x (x - 1)$. In this case, it is obvious that the constant functions $u_1(t) = 0$ and $u_2(t) = 1$ are two solutions of problem (1.5.1). Obviously, the nonlinear part is a locally Lipschitz function. So, the uniqueness result for initial value problems, due to Picard [28], does not hold in this situation.

Notice that the unique solution of problem

$$u'(t) = u(t) (u(t) - 1), \ t \in J, \quad u(a) = u_0 < 0$$

satisfies that $u'(t) > 0$ for all $t \in J$.

The same holds when $u_0 > 1$. On the contrary, if $u_0 \in (0, 1)$, we have that $u' < 0$ on J. In particular, we conclude that the set of solutions of (1.5.1) with $f(t, x) = x (x - 1)$ is reduced to $u_1 \equiv 0$ and $u_2 \equiv 1$. Since, for any $t_0 \in J$, the set

$$\{u(t_0), \ u \text{ is a solution of } u'(t) = u(t) (u(t) - 1), \ t \in J, \quad u(a) = u(b)\} = \{0, \ 1\}$$

is not connected and compact, we have constructed an example in which the Kneser's theorem [28] fails for boundary value problems.

These examples tell us that the existence of solutions for boundary value problems does not depend only on the regularity of the nonlinear part of the equation. In fact, the examples given above can be constructed because the operator $L u = u'$, which defines the linear part of the equation, is not invertible on the space of periodic functions $\{u \in \mathscr{C}^1(J, \mathbb{R}), \ u(a) = u(b)\}$. This property tells us that, what concerns in the existence of solutions of a nonlinear problem, the linear part of the equation, coupled with the imposed boundary conditions, plays a fundamental role. We remind that the invertibility of the linear operator is equivalent to the existence of a related Green's function.

In order to study the existence of a solution of problem (1.5.1), we may ask for an argument similar to Bolzano's lemma that, provided that a real continuous function attains different sign at the extremes of a given interval, ensures that it has a zero located in such interval. In our case, u is a solution of (1.5.1) if and only if

$$u'(t) - f(t, u(t)) = 0, \ t \in J, \quad u(a) - u(b) = 0.$$

So, the analogous statement to the Bolzano's lemma is written as follows:

Suppose that there are $\alpha, \beta \in \mathscr{C}^1(J, \mathbb{R})$, such that either $\alpha \leq \beta$ on J or $\alpha \geq \beta$ on J and that satisfy the following inequalities:

$$\alpha'(t) - f(t, \alpha(t)) \geq 0, \ t \in J, \quad \alpha(a) - \alpha(b) \geq 0$$

and

$$\beta'(t) - f(t, \beta(t)) \leq 0, \ t \in J, \quad \beta(a) - \beta(b) \leq 0.$$

Is it possible to ensure that there is $u \in \mathscr{C}^1(J, \mathbb{R})$, a solution of problem (1.5.1), lying between α and β?

The response is, in this case, affirmative. We present a proof which is a particular case of [5, Theorems 3.1 and 3.2]. An alternative proof can be found in [41].

We assume, in a first moment, that $\alpha \leq \beta$ on J.

The usual way to prove such assertion follows several steps:

Step 1: Construction of a modified problem

The first step of the proof consists of constructing a modified problem that must satisfy the following properties:

1. The nonlinear part of the modified boundary value problem must be continuous and bounded.
2. Both problems coincide on the sector

$$[\alpha, \beta] = \{u \in \mathscr{C}(J, \mathbb{R}), \ \text{such that} \ \alpha(t) \leq u(t) \leq \beta(t) \ \text{for all} \ t \in J\}.$$

To this end, we define the truncated function

$$p(t, x) = \max\{\alpha(t), \min\{x, \beta(t)\}\}.$$

It is obvious that this function is continuous and that for all $u \in \mathcal{C}(J, \mathbb{R})$ it is satisfied that

$$p(t, u(t)) = u(t) \quad \text{if and only if} \quad \alpha(t) \leq u(t) \leq \beta(t).$$

Now, choosing $m > 0$ arbitrary, we construct the following modified problem:

$$(P_m) \begin{cases} u'(t) - m\,u(t) = f(t, p(t, u(t))) - m\,p(t, u(t)), \\ \\ u(b) = p\,(b, u(b)) + p(a, u(a)) - p(b, u(b))). \end{cases}$$

From the definition of the function p, we have that problem (P_m) satisfies the two properties mentioned above.

Step 2: Every solution of the modified problem (P_m) is such that $\alpha(t) \leq u(t) \leq \beta(t)$ for all $t \in J$.

To deduce this result, we make use of the following maximum principle for terminal value problems:

Lemma 1.5.1. Let $M \in \mathbb{R}$ be given and $u \in \mathcal{C}^1([c, d], \mathbb{R})$. Suppose that

$$u'(t) + M\,u(t) \geq 0 \quad \text{for all } t \in [c, d], \qquad u(d) \leq 0,$$

then $u(t) \leq 0$ for all $t \in [c, d]$.

Proof. From the hypotheses, we have that there are $\sigma \in \mathcal{C}([c, d], \mathbb{R})$, $\sigma \geq 0$ in $[c, d]$ and $\mu \leq 0$ such that

$$u'(t) + M\,u(t) = \sigma(t) \quad \text{for all } t \in [c, d], \qquad u(d) = \mu.$$

Now, expressions (1.4.8) and (1.4.11) show us that this problem has a unique solution u which satisfies the following properties:

$$u(t) = -\int_t^d e^{-M\,(t-s)} \sigma(s)\, ds + e^{-M\,(t-d)} \mu \leq 0, \quad \text{for all } t \in [c, d],$$

and the proof is concluded. □

Returning to problem (P_m), we have, by the definition of p, that

$$\alpha(b) \leq u(b) \leq \beta(b).$$

In the case of $\alpha(c) > u(c)$ holds for some $c \in [a, b)$, then there exists $d \in (c, b]$ such that $\alpha(d) = u(d)$ with $\alpha(t) > u(t)$ for all $t \in [c, d)$. As a consequence, the definition of lower solution tells us

$$(\alpha - u)'(t) - m\,(\alpha - u)(t) \geq 0 \text{ for all } t \in [c, d] \text{ with } (\alpha - u)(d) = 0.$$

Now, Lemma 1.5.1 implies that $u \geq \alpha$ on $[c, d]$ and we arrive to a contradiction.

Analogously, we can prove that $u \leq \beta$ on J. So Step 2 is concluded.

Noticed that as an immediate consequence, we deduce that any solution, provided that it exists, of the modified problem (P_m) is a solution of problem (1.5.1) and it belongs to the sector $[\alpha, \beta]$.

To finish the existence result, we need to prove that the modified problem is solvable.

Step 3: There exists at least one solution of problem (P_m).

To deduce the solvability of the modified problem (P_m), we will use some classical results derived from fixed-point theorems related to operators defined in infinite dimensional spaces.

Definition 1.5.2 ([23, Definition 1.1, Appendix]). Let E and F be two real normed spaces, and $M \subset E$. The mapping $T : M \to F$ is called completely continuous if and only if it satisfies the two following properties:

1. T is continuous.
2. T maps bounded sets of M into relatively compact sets of F.

Example 1.5.3. Given a domain $\Omega \subset \mathbb{R}^n$, a Hilbert-Schmidt kernel is a function $k : \Omega \times \Omega \to \mathbb{R}$ such that

$$\int_\Omega \int_\Omega |k(t, s)|^2 \, dt \, ds < \infty.$$

The associated Hilbert-Schmidt integral operator $K : L^2(\Omega, \mathbb{R}) \to L^2(\Omega, \mathbb{R})$ is defined as

$$K\,u(t) = \int_\Omega k(t, s)\,u(s)\,ds.$$

One can see in [47] that this operator is completely continuous.

Example 1.5.4. Denoting by $\|u\|_k := \max\{\|u^{(j)}\|_\infty, \quad j = 0, \ldots, k\}$, from the Ascoli-Arzelà theorem [48], we have that the inclusion operator

$$i : (\mathscr{C}^{k+1}([a, b], \mathbb{R}), \|\cdot\|_{k+1}) \to (\mathscr{C}^k([a, b], \mathbb{R}), \|\cdot\|_k)$$

is completely continuous.

Now we enunciate the following fixed-point theorem due to Schaefer:

Theorem 1.5.5 ([33, Corollary 4.4.12]). *Let* X *be a normed space and* $T : X \to X$ *a completely continuous operator. Suppose that the set*

$$S = \{u \in X, \quad \text{for which there is some } \lambda \in [0, 1) \text{ such that } u = \lambda\, T\, u\}$$

is bounded in X.

Then, the operator T *has at least one fixed point.*

Notice that in an analogous way to Step 1, we have that the solutions of problem (P_m) coincide with the fixed points of the operator $\mathscr{T} : \mathscr{C}(J, \mathbb{R}) \to \mathscr{C}(J, \mathbb{R})$, defined as

$$\mathscr{T}\, u(t) = -\int_t^b e^{m\,(t-s)}\, \left(f(t, p(s, u(s))) - m\, p(s, u(s)) \right)\, ds$$

$$+ e^{m\,(t-b)}\, p\,(b, u(b) + p(a, u(a)) - p(b, u(b))), \quad \text{for all } t \in J.$$

It is not difficult to verify that this operator is completely continuous.

The definition of function p tells us that there is a constant $K > 0$ such that $\|\mathscr{T}\, u\|_\infty \le K$ for all $u \in \mathscr{C}(J, \mathbb{R})$. In particular, the set S introduced in Theorem 1.5.5 is bounded in $(\mathscr{C}(J, \mathbb{R}), \|\cdot\|_\infty)$ and operator $\mathscr{T}$ has at least one fixed point.

Conclusion: There exists a solution of problem (1.5.1) *lying between the lower solution* α *and the upper solution* β.

Indeed, the existence of a solution of problem (P_m) in $[\alpha, \beta]$, shown in Steps 2 and 3, implies, from Step 1, that the same property holds for problem (1.5.1).

It is important to point out that, on the contrary to problem (P_m), not all the solutions of problem (1.5.1) must be in $[\alpha, \beta]$. To see this, it is enough to consider the problem

$$u'(t) = u(t)\,(u(t) - 1), \ t \in J, \quad u(a) = u(b).$$

As we have noted previously, this problem has only two solutions, $u_1 \equiv 0$ and $u_2 \equiv 1$.

It is immediate to verify that $\alpha \equiv 1/2$ and $\beta \equiv 3/2$ are a pair of lower and upper solutions for this problem. Obviously, the constant solution u_2 belongs to the sector formed by this two functions, but $u_1 \notin [\alpha, \beta]$.

The case $\alpha \ge \beta$ in J can be treated in an analogous way. In this situation, instead of function p we must define

$$q(t, x) = \max\, \{\beta(t), \min\, \{x, \alpha(t)\}\},$$

and consider, for any $m > 0$, the modified problem

$$(Q_m)\ \begin{cases} u'(t) + m\, u(t) = f(t, q(t, u(t))) + m\, q(t, u(t)), \\ \qquad u(a) = q\,(b, u(a) + q(b, u(b)) - q(a, u(a))). \end{cases}$$

The proof follows similar steps and we need to apply the following anti-maximum principle for initial value problems. The proof of this result holds from (1.4.7) and (1.4.10).

Lemma 1.5.6. *Let $M \in \mathbb{R}$ be given and $u \in \mathscr{C}^1([c,d], \mathbb{R})$. Suppose that*

$$u'(t) + M\,u(t) \geq 0 \quad \text{for all } t \in [c,d], \qquad u(c) \geq 0,$$

then $u(t) \geq 0$ for all $t \in [c,d]$.

We can also consider the second-order periodic nonlinear boundary value problem

$$u''(t) = f(t, u(t)), \ t \in J, \quad u(a) = u(b), \ u'(a) = u'(b), \qquad (1.5.2)$$

and define a lower and an upper solution for this problem as

$$\alpha''(t) \geq f(t, \alpha(t)), \ t \in J, \quad \alpha(0) = \alpha(1), \ \alpha'(a) \geq \alpha'(b)$$

and

$$\beta''(t) \leq f(t, \beta(t)), \ t \in J, \quad \beta(0) = \beta(1), \ \beta'(a) \leq \beta'(b).$$

In this situation, one can prove that if there are $\alpha, \beta \in \mathscr{C}^2(J, \mathbb{R})$, such that $\alpha \leq \beta$ on J, then there is $u \in \mathscr{C}^2(J, \mathbb{R})$, a solution of problem (1.5.2), such that $u \in [\alpha, \beta]$.

The proof follows similar steps to the ones exposed above. The modified problem is defined, for $m > 0$ given, as

$$\begin{cases} u''(t) - m^2 u(t) = f(t, p(t, u(t))) - m^2 p(t, u(t)), & t \in J, \\ u(a) = u(b), \\ u'(a) = u'(b). \end{cases}$$

Step 2, in which it is proved that all the solutions of the modified problem belong to $[\alpha, \beta]$, follows from the fact that since

$$\alpha(a) - u(a) = \alpha(b) - u(b) = 0,$$

if there is $c \in (a, b)$ such that $\alpha(c) > u(c)$, then there exists $(t_0, t_1) \subset (a, b)$ such that

$$\alpha(t_0) - u(t_0) = \alpha(t_1) - u(t_1)$$

with $\alpha(t) > u(t)$ for all $t \in (t_0, t_1)$.

Moreover, from the definition of lower solution, we have that

$$(\alpha - u)''(t) - m^2 (\alpha - u)(t) \geq 0 \text{ for all } t \in [t_0, t_1].$$

The contradiction is attained by taking into account the following maximum principle for Dirichlet problems.

Lemma 1.5.7. *Let $m > 0$ be given and $u \in \mathscr{C}^2([c, d], \mathbb{R})$. Suppose that*

$$u'(t) - m^2 \, u(t) \geq 0 \quad \text{for all } t \in [c, d], \quad u(c) \leq 0, \; u(d) \leq 0.$$

Then $u(t) \leq 0$ for all $t \in [c, d]$.

Proof. The proof of this maximum principle holds from the fact that under these assumptions there are $\sigma \in \mathscr{C}([c, d], \mathbb{R})$, $\sigma \geq 0$ on $[c, d]$ and $\mu_1, \; \mu_2 \leq 0$ such that

$$u''(t) - m^2 \, u(t) = \sigma(t) \text{ for all } t \in [c, d], \qquad u(c) = \mu_1, \; u(d) = \mu_2.$$

Now, using the algorithm developed in [19] and shown in Appendix A of this book, we have that the function u follows the expression

$$u(t) = \int_c^d g_m(t, s) \, \sigma(s) \, ds$$

$$+ \frac{\sinh m \, (d - t)}{\sinh m \, (d - c)} \mu_1 + \frac{\sinh m \, (t - c)}{\sinh m \, (d - c)} \mu_2, \quad \text{for all } t \in [c, d],$$

with

$$g_m(t, s) = \frac{e^{-m(s+t)}}{2 \, m \, (e^{2dm} - e^{2cm})} \begin{cases} \left(e^{2cm} - e^{2ms}\right) \left(e^{2dm} - e^{2mt}\right), \text{ if } c \leq s \leq t \leq d, \\ \left(e^{2cm} - e^{2mt}\right) \left(e^{2dm} - e^{2ms}\right), \text{ if } c < t < s \leq d. \end{cases}$$

From the fact that Green's function $g_m \leq 0$ in $[c, d] \times [c, d]$ and the functions that multiply μ_1 and μ_2 are nonnegative on $[c, d]$, we conclude that $u \leq 0$ in $[c, d]$.   □

However, on the contrary to the first-order case, if α and β are given in the reversed order, i.e., $\alpha \geq \beta$ in J, the existence of solutions of the problem (1.5.2) cannot be ensured in general. Indeed, let us see the following problem:

$$u''(t) = -u(t) + \sin t, \; t \in [0, 2 \pi], \qquad u(0) = u(2 \pi), \; u'(0) = u'(2 \pi).$$

It is obvious that $\alpha \equiv 1$ and $\beta \equiv -1$ are a pair of reversed ordered lower and upper solutions of this problem. However, if we suppose that such problem has a solution u, then, multiplying both sides of the equation by $\sin t$ and using integration by parts, we arrive at the following contradiction:

$$\pi = \int_0^{2\pi} \sin^2 t \, dt = \int_0^{2\pi} u(t) \, \sin t \, dt + \int_0^{2\pi} u''(t) \, \sin t \, dt$$

$$= \int_0^{2\pi} u(t) \, \sin t \, dt - \int_0^{2\pi} u(t) \, \sin t \, dt = 0.$$

The main reason this happens is that if we try to make a parallel argument to the one given for the well-ordered case, to get a contradiction, it is necessary to have an anti-maximum principle of the type.

There is $m > 0$ such that for all $\sigma \in \mathscr{C}(J, \mathbb{R})$ with $\sigma \geq 0$ in J, if $u \in \mathscr{C}^2(J, \mathbb{R})$ is such that

$$u''(t) + m^2 u(t) = \sigma(t) \quad \text{for all } t \in J, \qquad u(a) = u(b) = 0, \tag{1.5.3}$$

then $u(t) \geq 0$ for all $t \in J$.

But this assertion does not hold for any positive m. Indeed, if u satisfies the previous assertion, then

$$u(t) = \int_a^b g_m(t, s) \sigma(s) \, ds,$$

with

$$g_m(t, s) = -\frac{1}{m \sin m (b-a)} \begin{cases} \sin m (s-a) \sin m (b-t), & \text{if } a \leq s \leq t \leq b, \\ \sin m (t-a) \sin m (b-s), & \text{if } a < t < s \leq b. \end{cases}$$

One can verify (see [14] for details) that this function is nonpositive on $J \times J$ if and only if $m \in (0, \pi/(b-a))$. Moreover, it changes sign for all $m > \pi/(b-a)$. We note that for $m = k \pi/(b-a)$, $k = 1, 2, \ldots$, Green's function is not defined.

Using this property, it is not difficult to construct [4], for any $m > 0$ given, a nonnegative continuous function σ for which the corresponding solution u is not positive on $[a, b]$. So the previous assertion is not true for arbitrary $\sigma \geq 0$.

It is important to mention that we have proved that the previous assertion is not true for some particular choices of σ. However, we can choose another nonnegative continuous function σ for which the unique solution u of problem (1.5.3) is nonnegative in J. This is the case of $\sigma = m^2$ that has as a unique solution

$$u(t) = 1 - \frac{\cos \frac{m (a+b-2t)}{2}}{\cos \frac{m (a-b)}{2}}.$$

It is immediate to verify that $u \leq 0$ in J if and only if $m \in (0, \pi/(b-a))$, $u \geq 0$ in J if and only if $m \in (\pi/(b-a), 2\pi/(b-a))$, and u changes its sign in J for all $m > 2\pi/(b-a)$.

As we have noted at the beginning of this section, the method of lower and upper solutions has been applied to more general and complicated situations: higher order differential equations, different boundary value conditions, dependence of the nonlinear part on the successive derivatives, functional dependence of the solutions, nonlinear boundary conditions, Carathéodory functions, discontinuities on the data of the equation, etc. We refer to [10] and the references therein for details.

1.6 Comparison Results

In the previous section we have shown some of the arguments used to ensure the existence of solutions by means of the method of lower and upper solutions. It has been pointed out in Lemmas 1.5.1, 1.5.6, and 1.5.7 that such existence results have a strong dependence on the comparison results of suitable linear operators.

This kind of comparison results represents a fundamental property of some linear operators: if the linear operator acting over a function has constant sign, then the considered function has also constant sign. As we have shown in the previous section, this is not a "universal" property. It is for this reason that, due to its application to nonlinear boundary value problems, the study of the operators that satisfy such property has a vital importance and has been extensively studied in the literature in the last decades.

In this section we will present some basic properties related to comparison principles of linear operators coupled with different classes of two-point boundary value conditions U_i, $i = 1, \ldots, n$, defined in (1.4.2). First, we define the following set for such a boundary condition:

$$X_U := \{u \in W^{n,1}(J, \mathbb{R}), \quad \text{such that } U_i(u) = 0 \text{ for all } i = 1, \ldots, n\}. \quad (1.6.1)$$

Now we introduce the concept of related set to a boundary condition. We recall that the choice of such set is not unique.

Definition 1.6.1. Let U_i, $i = 1, \ldots, n$, be the two-point boundary conditions defined in (1.4.2) and X_U be the space defined by (1.6.1). We say that $X(U) \subset W^{n,1}(J, \mathbb{R})$ is a related set to X_U, if it satisfies the following properties:

1. $X_U \subset X(U)$.

2. If $u, v \in X(U)$, then $\lambda u + \mu v \in X(U)$ for all $\lambda, \mu \geq 0$.

Next, we introduce the concept of inverse negative operator.

Definition 1.6.2. Let $X(U)$ be a related set to X_U, given by Definition 1.6.1. We say that the general nth-order linear differential operator L_n, defined in (1.4.3), is inverse negative on $X(U)$ if and only if for any $u \in X(U)$, the following property holds:

$$L_n u(t) \geq 0 \quad \text{for a.e. } t \in J \quad \text{implies that } u(t) \leq 0 \quad \text{for all } t \in J.$$

As an immediate consequence, [50, Proposition 1.1], we deduce the invertibility of the operator L_n on X_U.

Lemma 1.6.3. *If L_n is inverse negative on $X(U)$, then L_n is invertible on X_U.*

Proof. Let $u \in X_U$ be such that $L_n u = 0$. Since $X_U \subset X(U)$ we have that $u \leq 0$ on J. It is obvious that $-u \in X_U$ and $L_n(-u) = 0$; in consequence, $-u \leq 0$

on J. Therefore, $u \in X_U$ and $L_n u = 0$ if and only if $u \equiv 0$. This property is, by Theorem 1.4.2, equivalent to the invertibility of the linear operator L_n in X_U.   □

Example 1.6.4. If we consider, for instance, the second-order Dirichlet boundary conditions $u(0) = u(1) = 0$, we have that the functionals that characterize these boundary conditions are given as

$$U_1(u) = u(0) \quad \text{and} \quad U_2(u) = u(1).$$

So we have that the space defined in (1.6.1) is given by

$$X_U := \{u \in W^{2,1}([0,1], \mathbb{R}), \quad \text{such that } u(0) = u(1) = 0\}.$$

Of course, $X(U)$ can be defined as X_U, but we have, among others, different possibilities:

$$X(U) = \{u \in W^{2,1}(J, \mathbb{R}), \quad u(0) = 0, \quad u(1) \le 0\},$$
$$X(U) = \{u \in W^{2,1}(J, \mathbb{R}), \quad u(0) \le 0, \quad u(1) = 0\},$$

or

$$X(U) = \{u \in W^{2,1}(J, \mathbb{R}), \quad u(0) \le 0, \quad u(1) \le 0\}.$$

If we consider the nth-order periodic boundary conditions,

$$u^{(i)}(0) = u^{(i)}(1), \quad i = 0, \dots, n,$$

we have several possibilities to define the set $X(U)$, see for instance:

$$X(U) = \{u \in W^{n,1}(J, \mathbb{R}), \quad u^{(i)}(0) \ge u^{(i)}(1), \quad i = 0, \dots, n\},$$

$$X(U) = \{u \in W^{n,1}(J, \mathbb{R}), \quad u^{(i)}(0) = u^{(i)}(1), \quad i = 0, \dots, n-1, i \ne j, \quad u^{(j)}(0) \ge u^{(j)}(1)\}$$

or

$$X(U) = X_U = \{u \in W^{n,1}(J, \mathbb{R}), \quad u^{(i)}(0) = u^{(i)}(1), \quad i = 0, \dots, n\}.$$

Of course, the bigger the set $X(U)$ is, the more difficult it is for operator L_n to be inverse negative on $X(U)$.

First, we present the following relationship between the validity of a maximum principle and the sign of Green's function related to the problem

$$L_n u(t) = \sigma(t), \ t \in J, \quad U_i(u) = 0, \ i = 1, \dots, n. \tag{1.6.2}$$

Lemma 1.6.5. *Assume that problem* (1.6.2) *has a unique solution. If Green's function related to problem* (1.6.2) *has positive values at some point of its square of definition, then the operator* L_n *is not inverse negative on any set* $X(U)$ *related to* X_U.

Proof. First note that from Theorem 1.4.2 we know that there is a unique Green's function related to problem (1.6.2). The continuity of Green's function on the triangles $t < s$ and $s < t$ implies that if $g(t_0, s_0) > 0$ for some $(t_0, s_0) \in J \times J$, we can assume, without loss of generality, that $t_0 \neq s_0$. In consequence we know that there is $\varepsilon > 0$ such that $(s_0 - \varepsilon, s_0 + \varepsilon) \subset J$ for which $g(t_0, s) < 0$ for all $s \in (s_0 - \varepsilon, s_0 + \varepsilon)$.

In a similar way to [7, Theorem 2.2], we define the function

$$\sigma_0(s) = \begin{cases} \exp\left(-1/\left(1 - ((s - s_0)/\varepsilon)^2\right)\right), & \text{if } |s - s_0| < \epsilon \\ 0, & \text{if } |s - s_0| \geq \epsilon. \end{cases}$$

It is immediate to verify that $\sigma_0 \in \mathscr{C}^\infty(J, \mathbb{R})$ and $\sigma_0 \geq 0$ on J.

So, let u_0 be the unique solution of problem

$$L_n u(t) = \sigma_0(t), \ t \in J, \quad U_i(u) = 0, \ i = 1, \ldots, n.$$

By the definition of Green's function we have that

$$u_0(t_0) = \int_a^b g(t_0, s)\, \sigma_0(s)\, ds = \int_{s_0-\varepsilon}^{s_0+\varepsilon} g(t_0, s)\, \sigma_0(s)\, ds > 0.$$

In particular $L_n u_0 \geq 0$ in J but $u_0 \not\leq 0$ in J

Due to the fact that $U_i(u_0) = 0$ for all $i = 1, \ldots, n$, we know that $u_0 \in X_U \subset X(U)$. In consequence operator L_n is not inverse negative on $X(U)$. $\square$

As a direct consequence, we deduce the following characterization of the maximum principle in X_U.

Corollary 1.6.6. *The operator* L_n *is inverse negative on* X_U *if and only if Green's function related to problem* (1.6.2) *is nonpositive on its square of definition.*

It is important to point out that the nonpositiveness of Green's function is not sufficient to ensure the maximum principle in a more general set $X(U)$. In the next example we present a problem with negative Green's functions for which the operator is not inverse negative on $X(U)$.

Example 1.6.7. Consider the second-order operator $u'' + M u$ defined on the set

$$X(U) = \{u \in W^{2,1}(J, \mathbb{R}), \quad u(a) \geq u(b), \ u'(a) = u'(b)\}.$$

First, consider the problem

$$u''(t) + M u(t) = \sigma(t), \ t \in J, \quad u(a) = u(b), \ u'(a) = u'(b).$$

Using analogous arguments to the ones used in Example 1.4.3, one can verify that Green's function exists for all $M \neq ((2k\pi)/(b-a))^2$, $k = 0, 1, \ldots$, and has the following expression:

$$g_n(t,s) = \frac{-1}{2\sqrt{-M}\left(e^{b\sqrt{-M}} - e^{a\sqrt{-M}}\right)}$$

$$\times \begin{cases} e^{\sqrt{-M}(a-s+t)} + e^{\sqrt{-M}(b+s-t)}, & \text{if } a \leq s \leq t \leq b, \\ e^{\sqrt{-M}(a+s-t)} + e^{\sqrt{-M}(b-s+t)}, & \text{if } a \leq t < s \leq b, \end{cases}$$

if $M < 0$, and

$$g_p(t,s) = \frac{1}{2\sqrt{M}\sin\left(\frac{1}{2}\sqrt{M}(b-a)\right)}$$

$$\times \begin{cases} \cos\left(\frac{1}{2}\sqrt{M}(b-a+2s-2t)\right), & \text{if } a \leq s \leq t \leq b, \\ \cos\left(\frac{1}{2}\sqrt{M}(a-b+2s-2t)\right), & \text{if } a \leq t < s \leq b, \end{cases}$$

whenever $M > 0$.

In [32] it is shown that Green's function is negative on $J \times J$ if and only if $M < 0$. In consequence, as we have noted in Lemma 1.6.5, this operator cannot be inverse negative for all $M \geq 0$.

So, let $M < 0$ and $u \in X(U)$ be such that $u'' + M u \geq 0$ in J. Then there are $\sigma \geq 0$ in J and $\mu \geq 0$ such that

$$u''(t) + M u(t) = \sigma(t), \quad t \in J, \quad u(a) - u(b) = \mu, \quad u'(a) = u'(b),$$

or, which is the same,

$$u(t) = \int_a^b g_n(t,s)\,\sigma(s)\,ds - \frac{e^{\sqrt{-M}t} - e^{\sqrt{-M}(a+b-t)}}{2\left(e^{b\sqrt{-M}} - e^{a\sqrt{-M}}\right)}\mu$$

Now, it is enough to consider $\sigma \equiv 0$ and $\mu = 1$ to deduce that the unique solution u changes its sign on J. In consequence, despite Green's function being nonpositive for all $M < 0$, the operator $u'' + M u$ is not inverse negative in $X(U)$ for all $M \in \mathbb{R}$.

From Corollary 1.6.6 we know that the inverse negative character of the operator L_n in X_U is equivalent to the nonpositiveness of Green's function in $J \times J$. One can think that this is the only situation in which such property occurs. However, this is not true at all. As we show in the next example, in some particular situations, the sign of Green's function can characterize the inverse negative property in a bigger set of solutions than X_U.

Example 1.6.8. Consider the space

$$X(U) = \{u \in W^{n,1}(J, \mathbb{R}), \quad u^{(i)}(a) = u^{(i)}(b), \ i = 0, \ldots, n - 2, \ u^{(n-1)}(a) \geq u^{(n-1)}(b) \}.$$

Suppose that the general nth-order linear differential operator L_n, defined by (1.4.3), is inverse negative on $X(U)$. From Lemma 1.6.3, we have that there is L_n^{-1} defined on the space

$$X_U = \{u \in W^{n,1}(J, \mathbb{R}), \quad u^{(i)}(a) = u^{(i)}(b), \ i = 0, \ldots, n - 1\}. \qquad (1.6.3)$$

Theorem 1.4.2 ensures that there is a unique Green's function g related to this problem.

So, let $u \in X(U)$ be such that $L_n u \geq 0$ in J. Then there are $\sigma \geq 0$ on J and $\mu \geq 0$, satisfying

$$L_n u(t) = \sigma(t), \ t \in J, \quad u^{(i)}(a) = u^{(i)}(b), \ i = 0, \ldots, n - 2, \ u^{(n-1)}(a) = u^{(n-1)}(b) = \mu.$$

Expression (1.4.9) implies that the function u follows the expression

$$u(t) = \int_a^b g(t, s) \, \sigma(s) \, ds + g(t, a) \, \mu.$$

From this expression, it is obvious that operator L_n is inverse negative on $X(U)$ if and only if $g \leq 0$ on $J \times J$. But this property is, from Corollary 1.6.6, equivalent to the fact that L_n is inverse negative on X_U.

In an analogous way to the previous situations, we can define an inverse positive operator as follows.

Definition 1.6.9. Let $X(U)$ be a related set to X_U, given in Definition 1.6.1. We say that the general nth-order linear differential operator L_n, defined in (1.4.3), is inverse positive on $X(U)$ if and only if for all $u \in X(U)$, the following property holds:

$$L_n u(t) \geq 0 \ \text{ for a.e. } t \in J \quad \text{implies that } u(t) \geq 0 \quad \text{for all } t \in J.$$

As in Lemma 1.6.3 we have that it is invertible on X_U.

Lemma 1.6.10. *If L_n is inverse positive on $X(U)$, then L_n is invertible on X_U.*

The following result can be proved in a similar way to Lemma 1.6.5.

Lemma 1.6.11. *If Green's function related to problem (1.6.2) attains negative values at some point of its square of definition, then the operator L_n is not inverse positive on any set $X(U)$ related to X_U.*

So, we arrive at the following consequence.

Corollary 1.6.12. *The operator L_n is inverse positive on X_U if and only if Green's function related to problem* (1.6.2) *is nonnegative on its square of definition.*

As in Example 1.6.4, we can construct different possibilities of related sets $X(U)$.

Now, in a parallel way to Example 1.6.7, we present an example for which the nonnegativeness of Green's function is not sufficient to ensure the anti-maximum principle in a more general set $X(U)$.

Example 1.6.13. Consider the second-order operator $u'' + M u$ defined on the set

$$X(U) = \{u \in W^{2,1}(J, \mathbb{R}), \quad u(a) \geq u(b), \ u'(a) = u'(b)\}.$$

As we have shown in Example 1.6.7, we know that the corresponding Green's function is given by the expressions g_n (when $M < 0$) and g_p (if $M > 0$ and $M \neq ((2 k \pi)/(b - a))^2, k = 0, 1, \ldots)$.

In [4] it is shown that Green's function is nonnegative on $J \times J$ if and only if $M \in (0, (\pi/(b - a))^2]$. In consequence, as we have noted in Lemma 1.6.5, this operator cannot be inverse positive for all $M \in (-\infty, 0] \cup ((\pi/(b - a))^2, \infty)$.

So, let $M \in (0, (\pi/(b - a))^2]$ and $u \in X(U)$ be such that $u'' + M u \geq 0$ in J. Then there are $\sigma \geq 0$ in J and $\mu \geq 0$ such that

$$u''(t) + M u(t) = \sigma(t), \ t \in J, \quad u(a) - u(b) = \mu, \ u'(a) = u'(b),$$

or, which is the same,

$$u(t) = \int_a^b g_p(t, s)\, \sigma(s)\, ds + \frac{\sin\left(\frac{\sqrt{M}(a+b-2t)}{2}\right)}{2 \sin\left(\frac{\sqrt{M}(b-a)}{2}\right)} \mu.$$

Now, choosing $\sigma \equiv 0$ and $\mu = 1$ we conclude that the unique solution u changes its sign on J. In consequence, despite Green's function being nonnegative for all $M \in (0, (\pi/(b - a))^2]$, there is no $M \in \mathbb{R}$ for which the operator $u'' + M u$ is inverse positive in $X(U)$.

In a similar way to Example 1.6.8, the sign of Green's function can characterize the inverse positive property in a bigger set of solutions than X_U. Let us see the following example.

Example 1.6.14. From expressions (1.4.7) and (1.4.10), we can deduce that the general nth-order linear differential operator L_n is inverse positive on

$$X_U = \{u \in W^{n,1}(J, \mathbb{R}), \quad u^{(i)}(a) = 0, \ i = 0, \ldots, n - 1\}$$

if and only if it is inverse positive on

$$X(U) = \{u \in W^{n,1}(J, \mathbb{R}), \quad u^{(i)}(a) = 0, \ i = 0, \ldots, n - 2, \ u^{(n-1)}(a) \geq 0\}.$$

It is important to point out that, by definition of $X(U)$, L_n cannot be inverse negative on $X(U)$. This property tell us that the same holds in X_U. We note that we have not this information by using only the definition of X_U.

On the other hand, from (1.4.8) and (1.4.11), we obtain that L_n is inverse positive (negative) on

$$X_U = \{u \in W^{n,1}(J, \mathbb{R}), \quad u^{(i)}(b) = 0, \ i = 0, \ldots, n-1 \}$$

if and only if it is inverse positive (negative) on

$$X(U) = \{u \in W^{n,1}(J, \mathbb{R}), \quad u^{(i)}(b) = 0, \ i = 0, \ldots, n-2, \ u^{(n-1)}(b) \leq 0 \}.$$

In this case, the definition of $X(U)$ implies that L_n cannot be inverse negative on $X(U)$ if n is even and it cannot be inverse positive on $X(U)$ when n is odd. As consequence the same property holds in X_U. As in the initial case, we are not able to deduce this property only with the definition of X_U.

To finish this section, note that, as a direct consequence of Corollaries 1.6.6 and 1.6.12 and the expressions (1.4.13) and (1.4.14), we obtain the following relationship between the inverse positive character of a linear operator and its adjoint.

Theorem 1.6.15. *Let L_n be the general nth-order linear operator defined in (1.4.3) and U_i, $i = 1, \ldots, n$, be the general two-point boundary conditions defined in (1.4.2). Then the following equivalence holds:*
Operator L_n is inverse positive (negative) on $D(L) = X_U$ if and only if its adjoint operator L_n^* is inverse positive (negative) on $D(L_n^*)$.

As a particular case, we have that if L_n is the general nth-order linear operator defined in (1.4.3) with $a_k \in \mathscr{C}^{n-k}(J, \mathbb{R})$. The operator L_n is inverse positive (negative) on $D(L) = X_U$ if and only if its adjoint operator L_n^* defined by (1.4.15) is inverse positive (negative) on $D(L_n^*)$ given by (1.4.16).

Corollary 1.6.16. *Suppose that the linear operator L_n defined in (1.4.3) has constant coefficients. Then it is inverse positive (negative) on the set*

$$X(U) = \{u \in W^{n,1}(J, \mathbb{R}), \quad u^{(i)}(a) = u^{(i)}(b), \ i = 0, \ldots, n-2, \ u^{(n-1)}(a) \geq u^{(n-1)}(b) \}$$

if and only if the operator

$$\hat{L}_n u(t) = u^{(n)}(t) - a_1 u^{(n-1)}(t) + \cdots + (-1)^{n-1} a_{n-1} u'(t) + (-1)^n a_n u(t)$$

is inverse negative (positive) on $X(U)$ if n is odd, or inverse positive (negative) on $X(U)$ if n is even.

Proof. First, notice that, as we have shown in Example 1.6.8, the operator L_n is inverse negative on $X(U)$ if and only if $g(t, s) \leq 0$ for all $(t, s) \in J \times J$ and $\hat{L}_n$ is inverse positive on $X(U)$ if and only if $\hat{g}(t, s) \geq 0$ for all $(t, s) \in J \times J$.

As we have seen in Sect. 1.4, $L_n^* = (-1)^n \hat{L}_n$. So (1.4.13) and (1.4.14) tell us that $D(L^*) = X_U$ and that the corresponding Green's functions g and $\hat{g}$ satisfy the equality

$$g(t, s) = (-1)^n \hat{g}(s, t).$$

$\square$

In an analogous way, making use of Example 1.3.3, we arrive at the following result.

Corollary 1.6.17. *Suppose that the linear operator* L_n *defined in (1.4.3) has constant coefficients. Then it is inverse positive on the set*

$$\{u \in W^{n,1}(J, \mathbb{R}), \quad u^{(i)}(a) = 0, \ i = 0, \ldots, n-2, \ u^{(n-1)}(a) \geq 0 \}$$

if and only if the operator

$$\hat{L}_n u(t) = u^{(n)}(t) - a_1 u^{(n-1)}(t) + \cdots + (-1)^{n-1} a_{n-1} u'(t) + (-1)^n a_n u(t)$$

is inverse negative, if n is odd, or inverse positive, if n is even, on the set

$$\{u \in W^{n,1}(J, \mathbb{R}), \quad u^{(i)}(b) = 0, \ i = 0, \ldots, n-2, \ u^{(n-1)}(b) \leq 0 \}.$$

1.7 Monotone Iterative Techniques

As we have seen in the previous section, a comparison result in X_U holds if and only if the related Green's function has constant sign on $J \times J$. In this section we will use this property to obtain the exact expression of some of the solutions that are located between the lower solution α and the upper solution β of a nonlinear problem, provided that they exist. Such expression will be obtained by means of the construction of two monotone sequences that start at α and β and that are given as the unique solutions of linear problems related to the nonlinear equation. We point out that this process is, in some sense, a linearization of the studied equation and the uniqueness of solution of the linear problems is equivalent to the existence of the related Green's functions. As in the case of the lower and upper solutions, this is a tool used for nonlinear boundary value problems, but we will introduce it here because it will be fundamental in the next section, in which a one parameter family of nth-order linear operators is studied.

Next, we present an example in which this method is developed. It is a generalization of the work done in [7] for the periodic case. To be concise, we consider the following nonlinear boundary value problem:

$$L_n u(t) = f(t, u(t)) \text{ for a.e. } t \in J, \quad U_i(u) = 0, \ i = 1, \ldots, n. \qquad (1.7.1)$$

Here L_n is the general nth-order linear operator defined in (1.4.3) and U_i, $i =$
$1, \ldots, n$, are defined in (1.4.2) and they represents the general two-point linear
boundary conditions.

The nonlinear part $f : J \times \mathbb{R} \to \mathbb{R}$ is assumed to be a Carathéodry function, i.e.,
$f(\cdot, x)$ is measurable for all $x \in \mathbb{R}$, $f(t, \cdot)$ is continuous for a. e. $t \in J$, and, for
every $R > 0$, there exists $h_R \in L^1(J, \mathbb{R})$ such that:

$$| f(t, x) | \leq h_R(t) \text{ for a. e. } t \in J$$

with $\|x\|_\infty \leq R$.

One can see in [2, Theorem 3.1] that if f is a Carathéodory function and
$u \in \mathcal{L}^1(J, \mathbb{R})$, then the superposition operator $f(\cdot, u(\cdot)) \in \mathcal{L}^1(J, \mathbb{R})$. As a
consequence, we are looking for solutions u belonging to the space X_U introduced
in (1.6.1).

Now, we define the concept of lower and upper solutions for problem (1.7.1) as
follows.

Definition 1.7.1. Let $\alpha \in X(U)$, with $X(U)$ a related set to X_U introduced in
Definition 1.6.1. We say that α is a lower solution of the problem (1.7.1) if α satisfies

$$L_n \alpha(t) \geq f(t, \alpha(t)) \text{ for a.e. } t \in J.$$

Analogously, β is an upper solution of the problem (1.7.1) if $-\beta \in X(U)$ and it
satisfies

$$L_n \beta(t) \leq f(t, \beta(t)) \text{ for a.e. } t \in J.$$

Define now, for every $M \in \mathbb{R}$, the nth-order linear differential operator

$$T_n[M] u(t) = L_n u(t) + M u(t), \quad t \in J. \tag{1.7.2}$$

The following condition is assumed on function f:

(H_d) There is $M \in \mathbb{R}$, such that $f(t, x) + M x \leq f(t, y) + M y$ for a.e. $t \in J$
and $\alpha(t) \leq y \leq x \leq \beta(t)$.

Now, we are in a position to prove the following result which is known as the
monotone iterative technique. It is essentially [7, Theorem 2.1].

Theorem 1.7.2. *Suppose that there exist $\alpha \leq \beta$ a pair of well-ordered lower and
upper solutions respectively for the nth-order nonlinear boundary value problem
(1.7.1). Assume that the function f satisfies the condition (H_d) for some $M \in \mathbb{R}$
such that the operator $T_n[M]$ is inverse negative in $X(U)$.*

*Then there exist two monotone sequences $\{\beta_m\}$ and $\{\alpha_m\}$, nonincreasing and
nondecreasing respectively, with $\beta_0 = \beta$ and $\alpha_0 = \alpha$, which converge uniformly to
the extremal solutions in $[\alpha, \beta]$ of the problem (1.7.1).*

Proof. Let $\eta \in \mathscr{L}^1(J, \mathbb{R})$ be such that $\alpha(t) \le \eta(t) \le \beta(t)$ for a.e. $t \in J$. Consider the following linear problem:

$$T_n[M] u(t) = f(t, \eta(t)) + M \eta(t), \quad t \in J, \quad U_i(u) = 0, \ i = 1, \ldots, n. \quad (1.7.3)$$

From Lemma 1.6.3, we know that the inverse negative character of operator $T_n[M]$ in $X(U)$ implies that this problem has a unique solution.

Using the condition (H_d) we have that $T_n[M](u - \beta) \ge 0$ a.e. on J. Since $u - \beta \in X(U)$ and $T_n[M]$ is inverse negative in this set we have that $u \le \beta$ on J.

Analogously we can prove that $\alpha \le u$ on J.

Let now u_1 and u_2 be the unique solutions of (1.7.3) for $\eta = \eta_1$ and $\eta = \eta_2$ respectively. Obviously, $u_1 - u_2 \in X_U$.

Thus, for $\alpha \le \eta_1 \le \eta_2 \le \beta$ we have that $T_n[M](u_1 - u_2) \ge 0$ a.e. on J, which implies that $u_1 \le u_2$ on J.

The sequences $\{\beta_m\}$ and $\{\alpha_m\}$ are obtained by recurrence: $\beta_0 = \beta$, $\alpha_0 = \alpha$, and for $m \ge 1$:

$$T_n[M] \beta_m(t) = f(t, \beta_{m-1}(t)) + M\beta_{m-1}(t), \quad U_i(\beta_m) = 0, \ i = 1, \ldots, n.$$

and

$$T_n[M] \alpha_m(t) = f(t, \alpha_{m-1}(t)) + M\alpha_{m-1}(t), \quad U_i(\alpha_m) = 0, \ i = 1, \ldots, n.$$

From the properties of Green's function related to the operator $T_n[M]$, it is immediate to verify that the sequences $\{\alpha_m\}$ and $\{\beta_m\}$ are bounded in the space $W^{n,1}(J, \mathbb{R})$. Thus, by standard arguments [32], we obtain that $\{\beta_m\}$ and $\{\alpha_m\}$ converge uniformly to Φ and ϕ, respectively. Moreover, such limit functions are a pair of solutions of problem (1.7.1).

Furthermore, if $x \in [\alpha, \beta]$ is a solution of (1.7.1), then $\alpha_m \le x \le \beta_m$, for all $m \in \mathbb{N}$. In consequence, passing to the limit in m, we have that $\phi \le x \le \Phi$, i. e., both functions are the extremal solutions of problem (1.7.1) in $[\alpha, \beta]$. □

We note that condition (H_d) is equivalent to imposing that the function $f(t, x) + M x$ decreases in $x \in [\alpha(t), \beta(t)]$. Clearly, the smaller the value of M is, the weaker the condition on the function f gets.

It is not assumed the negativeness of M because there are a lot of operators $T_n[M]$ for which the operator is inverse negative and $M > 0$. This is the case, for instance, of the operator $u'' + M u$: it is very well known [14] that it is inverse negative for all $M \in (-\infty, (\pi/(b - a))^2)$ on the set

$$\{u \in W^{2,1}(J, \mathbb{R}), \quad u(a) \le 0, \ u(b) \le 0\}.$$

Remark 1.7.3. Notice that, from Example 1.6.8, if we study the nth-order periodic boundary value problem

$$L_n u(t) = f(t, u(t)) \text{ for a.e. } t \in J, \quad u^{(i)}(a) = u^{(i)}(b), \ i = 0, \ldots, n-1,$$

we can define a lower solution and an upper solution as follows:

$$\begin{cases} L_n \alpha(t) \geq f(t, \alpha(t)) \text{ for a.e. } t \in J, \\ \alpha^{(i)}(a) = \alpha^{(i)}(b), \ i = 0, \ldots, n-2, \\ \alpha^{(n-1)}(a) \geq \alpha^{(n-1)}(b) \end{cases}$$

and

$$\begin{cases} L_n \beta(t) \leq f(t, \beta(t)) \text{ for a.e. } t \in J, \\ \beta^{(i)}(a) = \beta^{(i)}(b), \ i = 0, \ldots, n-2, \\ \beta^{(n-1)}(a) \leq \beta^{(n-1)}(b), \end{cases}$$

and we deduce existence and approximation results under weaker assumptions than assuming the equalities in the $(n-1)$th-derivatives.

Now, we may ask about the possibility of ensuring the existence and approximation of solutions lying between a pair of a lower and an upper solution given in reversed order, i.e., $\alpha \geq \beta$ in J. In this new situation, we must impose a parallel condition to (H_d). Now, the nondecreasing character is needed.

(H_i) There is $M \in \mathbb{R}$, such that $f(t, x) + M x \geq f(t, y) + M y$ for a.e. $t \in J$ and $\beta(t) \leq y \leq x \leq \alpha(t)$.

Following the same steps as in Theorem 1.7.2, we can prove the following result.

Theorem 1.7.4. *Suppose that there exist $\beta \leq \alpha$, a pair of reversed ordered upper and lower solutions, respectively, for the nth-order nonlinear boundary value problem (1.7.1). Assume that the function f satisfies the condition (H_i) for some $M \in \mathbb{R}$ such that the operator $T_n[M]$ is inverse positive in $X(U)$.*

Then there exist two monotone sequences $\{\beta_m\}$ and $\{\alpha_m\}$, nondecreasing and nonincreasing, respectively, with $\beta_0 = \beta$ and $\alpha_0 = \alpha$, which converge uniformly to the extremal solutions in $[\beta, \alpha]$ of the problem (1.7.1).

We note that in this case the bigger is the value of M, the weaker is the condition (H_i).

In this case, we do not assume that $M > 0$ because there are a lot of operators $T_n[M]$ for which the operator is inverse positive and $M < 0$. This is the case, for instance, of operator $u' + M\,u$ on the set

$$\{u \in W^{1,1}(J, \mathbb{R}), \quad u(a) \geq 0\},$$

which is inverse positive for all $M \in \mathbb{R}$.

1.8   Parameter Set of Constant Sign Green's Functions

This section is devoted to the study of the behavior of Green's function related to the operator

$$T_n[M] : X_U \to \mathscr{L}^1(J, \mathbb{R}),$$

defined in (1.4.3) and (1.7.2), with respect to the real parameter M.

First, we deduce a monotonicity dependence when the corresponding Green's functions have constant sign.

We note that this comparison result follows from Theorems 1.7.2 and 1.7.4, in which the monotone method is developed. This property gives us a new relationship between the lower and upper solution method and the constant sign properties of Green's function. We remark that the validity of the iterative techniques developed in those results hold if a maximum or an anti-maximum principle is fulfilled.

The result is proved below; the used arguments are similar to the ones given in [11, Lemma 2.8] where a stronger comparison result if obtained for the second-order Hill's equation.

Theorem 1.8.1. *Let M_1, $M_2 \in \mathbb{R}$ and suppose that the nth-order linear two-point boundary value problem*

$$T_n[M]\,u(t) = \sigma(t), \ t \in J, \quad U_i(u) = 0, \ i = 1, \ldots, n. \qquad (1.8.1)$$

has a unique solution for $M = M_j$, $j = 1, 2$. Let g_j be Green's functions related to the operator $T_n[M_j]$ and suppose that $g_j(t, s) \leq 0$ for all $(t, s) \in J \times J$ and $j = 1, 2$. Then, if $M_1 < M_2$, it is satisfied that $g_2 \leq g_1 \leq 0$ on $J \times J$.

Proof. Fix $\sigma \geq 0$ in J and denote as u_j the unique solution of problem (1.8.1) with $M = M_j$, $j = 1, 2$.

Consider now the following problem:

$$L_n\,u(t) = -M_1\,u(t) + \sigma(t), \ t \in J, \quad U_i(u) = 0, \ i = 1, \ldots, n. \qquad (1.8.2)$$

Since it is equivalent to (1.8.1) for $M = M_1$, we know that it has u_1 as its unique solution.

It is obvious that $\beta \equiv 0 \in X_U$ is an upper solution of this problem.

On the other hand, since, $g_2 \leq 0$ on $J \times J$, we have that $u_2 \leq 0$ on J. In consequence

$$L_n u_2(t) = -M_2 u_2(t) + \sigma(t) \geq -M_1 u_2(t) + \sigma(t), \ t \in J, \quad U_i(u_2) = 0, \ i = 1, \ldots, n,$$

or, which is the same, $u_2 \in X_U$ is a lower solution of the considered problem.

Note that problem (1.8.2) is a particular case of (1.7.1), with $f(t, x) + M_1 x = \sigma(t)$, which is a nonincreasing function in x for any fixed $t \in J$.

Corollary 1.6.6 tells us that $g_1 \leq 0$ on $J \times J$ is equivalent to the fact that the operator $T_n[M_1]$ is inverse negative in X_U. As a consequence, condition (H_d) is fulfilled and Theorem 1.7.2 implies that problem (1.8.2) has at least one solution $v \in [u_2, 0]$. The invertibility of operator $T_n[M_1]$ in X_U tells us that $v = u_1$ is unique.

Now, since the equality

$$\int_a^b (g_2(t, s) - g_1(t, s)) \sigma(s) \, ds = (u_2 - u_1)(t) \leq 0 \quad \text{for all } t \in J,$$

is valid for any nonnegative $\sigma \in \mathscr{L}^1(J, \mathbb{R})$, we conclude that $g_2(t, s) \leq g_1(t, s)$ for all $(t, s) \in J \times J$.   □

The previous result can be improved in the sense that both functions g_1 and g_2 cannot coincide in a suitable subset of $J \times J$.

Lemma 1.8.2. *Under the hypotheses of Theorem 1.8.1 the following properties hold:*

(i) *For all $t_0 \in (a, b)$ and every $\varepsilon > 0$ there is $t \in J_0 = (t_0 - \varepsilon, t_0 + \varepsilon)$ and an interval $J_t \subset J$ such that $0 \geq g_1(t, s) > g_2(t, s)$ for all $s \in J_t$.*

(ii) *If $g_1(t_0, s_0) < 0$ for some $(t_0, s_0) \in J \times J$, then there exists $r > 0$ such that $0 > g_1(t, s_0) > g_2(t, s_0)$ for a.e. $t \in (t_0 - r, t_0 + r)$.*

Proof. To prove (i), assume, on the contrary, that for some $t_0 \in J$ and $\varepsilon > 0$ it is satisfied that $g_1(t, s) = g_2(t, s)$ for all $(t, s) \in J_0 \times J$.

Fix $\sigma \geq 0$ in J, such that $\sigma > 0$ in J_0. Let u_j be the unique solution of problem (1.8.1) with $M = M_j$, $j = 1, 2$ for such a σ.

First, note that if $u_2 \equiv 0$ in J_0, then

$$0 = T_n[M_2] u_2(t) = \sigma(t) > 0, \quad \text{for all } t \in J_0,$$

and we arrive to a contradiction.

As a consequence, since $g_2 \leq 0$ on $J \times J$, we have that there is an interval $J_\epsilon \subset J_0$ such that $u_2 < 0$ on J_ϵ.

Now, by definition of u_1 and u_2, we have that for all $t \in J$ the following equality holds:

$$0 = L_n (u_1 - u_2)(t) + M_1 u_1(t) - M_2 u_2(t) = T_n[M_1] (u_1 - u_2)(t) + (M_1 - M_2) u_2(t).$$

If $g_1(t, s) = g_2(t, s)$ for all $(t, s) \in J_0 \times J$, we have that $u_1(t) = u_2(t)$ for all $t \in J_0$. Therefore $T_n[M_1] (u_1 - u_2) = 0$ in J_0.

The contradiction follows from the fact that $(M_1 - M_2) u_2(t) > 0$ for all $t \in J_\epsilon \subset J_0$.

To verify assertion (ii), we can assume, without loss of generality, that $a < t_0 < s_0 < b$. Arguing as in the previous case, taking into account condition (g4) in Definition 1.4.1, we deduce that

$$T_n[M_2] (g_1 - g_2)(t, s_0) = (M_2 - M_1) g_1(t, s_0), \quad \text{for a. e. } t \in (a, s_0). \quad (1.8.3)$$

Now, the continuity of function $g_1(\cdot, s_0)$ in (a, s_0) implies that it remains strictly negative for all $t \in (t_0 - r, t_0 + r)$, for some $r > 0$. As a consequence we have that $(g_1 - g_2)(t, s_0)$ cannot vanish a.e. $t \in (t_0 - r, t_0 + r)$. □

In the previous result it is not proved that functions g_1 and g_2 differ in the whole interval $(t_0 - r, t_0 + r)$. As we will see now, this property is true if $n \leq 2$ and the coefficient functions in L_n are continuous.

Corollary 1.8.3. *Under the hypotheses of Theorem 1.8.1, assuming that $n \leq 2$ and that the coefficient functions in the general nth-order linear operator L_n defined in (1.4.3) are continuous, we have that if $g_1(t_0, s_0) < 0$ for some $(t_0, s_0) \in J \times J$, then there exists $r > 0$ such that $0 > g_1(t, s_0) > g_2(t, s_0)$ for all $t \in (t_0 - r, t_0 + r)$*

Proof. First, note that the continuity assumption on the coefficients of operator L_n implies that equality (1.8.3) is fulfilled for all $t \in (a, s_0)$. Now, since $g_1 \geq g_2$ in $J \times J$, if $g_1(t_1, s_0) = g_2(t_1, s_0)$ for some $t_1 \in (t_0 - r, t_0 + r)$, we have that

$$T_n[M_2] (g_1 - g_2)(t_1, s_0) = 0, \quad \text{if } n = 1,$$

and

$$T_n[M_2] (g_1 - g_2)(t_1, s_0) = \frac{\partial^2}{\partial t^2}(g_1 - g_2)(t_1, s_0) \geq 0, \quad \text{if } n = 2,$$

and we arrive to a contradiction. □

We note that there is the possibility of getting some $t_1 \in J$ for which $g_2(t_1, s) = g_1(t_1, s)$ for all $s \in J$. To see this, it suffices to take into account the Dirichlet conditions $u(a) = u(b) = 0$. From condition (g6) in Definition 1.4.1, we have that the related Green's function satisfies $g(a, s) = g(b, s) = 0$ for all $s \in J$ and every real parameter M for which such function exists.

Even in the case of $g_1 \geq g_2$ on $J \times J$ but $g_1 \not\equiv g_2$ on $J \times J$, it is not ensured the existence of some $t_1 \in J$ satisfying $g_1(t_1, s) > g_2(t_1, s)$ for all $s \in J$. This is the case of the initial value problem, which, as we have remarked in (1.4.10), satisfies that its related Green's function is such that $g(a, s) = 0$ for all $s \in J$ and $g(t, s) = 0$ for all $s \in (t, b]$.

Despite this, in some cases, we can ensure that $g_1(t, s) > g_2(t, s)$ for all $(t, s) \in J \times J$. This is the situation of the second-order operator $u'' + M u$ with periodic boundary value conditions, see [11, Lemma 2.8] for details.

The proof of Theorem 1.8.1 has been done in the framework of the lower and upper solutions method coupled with the monotone iterative techniques pointed out in the two previous sections. As we will see now, the proof of this result can be made in a more general setting concerned with monotone operators defined in partially ordered sets. The result is the following.

Lemma 1.8.4. *Let X, Y be two nonempty partially ordered sets, and let $T_1, T_2 :$ $X \longrightarrow Y$ be two injective mappings such that*

$$T_1(x) \leq T_2(x) \quad \text{for all } x \in X.$$

Then, if T_1^{-1} or T_2^{-1} is nonincreasing (respectively, nondecreasing) on Y, we have that for all $y \in T_1(X) \cap T_2(X)$ it is satisfied that

$$T_1^{-1}(y) \leq T_2^{-1}(y) \ (\text{respectively,} \ T_1^{-1}(y) \geq T_2^{-1}(y)).$$

Proof. Let us consider first the case in which T_1^{-1} is nonincreasing. Let $y \in T_1(X) \cap T_2(X)$ be fixed; we have

$$T_1(T_2^{-1}(y)) \leq T_2(T_2^{-1}(y)) = y = T_1(T_1^{-1}(y)),$$

and then $T_2^{-1}(y) \geq T_1^{-1}(y)$ because T_1^{-1} is nonincreasing.

If we assume instead that T_2^{-1} is nonincreasing, then for each $y \in T_1(X) \cap T_2(X)$ we write

$$T_2(T_2^{-1}(y)) = y = T_1(T_1^{-1}(y)) \leq T_2(T_1^{-1}(y)),$$

which implies $T_2^{-1}(y) \geq T_2^{-1}(y)$.

The proofs when T_1^{-1} or T_2^{-1} is nondecreasing are similar. $\square$

As a direct consequence of this result we are in a position to give an alternative proof of Theorem 1.8.1 as follows.

Proof of Theorem 1.8.1. Consider the sets $X = \{u \in X_U, \ u \leq 0 \text{ on } J\}$ and $Y = \mathscr{L}^1(J, \mathbb{R})$ equipped with their usual pointwise partial orderings.

The mappings $T_n[M_j] : X \longrightarrow Y \ (j = 1, 2)$ are injective, and

$$\tilde{Y} = \{\sigma \in Y, \ \sigma \geq 0 \text{ a.e. in } J\} \subset T_n[M_1](X) \cap T_n[M_2](X).$$

Moreover, for every $u \in X$ we have, since $u \leq 0$ on J, that

$$T_n[M_2]u \leq T_n[M_1]u.$$

Finally, the inverse mappings $T_n^{-1}[M_j] \ (j = 1, 2)$ are nonincreasing, so Lemma 1.8.4 ensures that for every $\sigma \in \tilde{Y}$ and all $t \in J$ we have

$$\int_a^b g_2(t,s)\sigma(s)\,ds = T_n^{-1}[M_2]\sigma(t)$$

$$\leq T_n^{-1}[M_1]\sigma(t) = \int_a^b g_1(t,s)\sigma(s)\,ds,$$

hence $g_2 \leq g_1$ on $J \times J$.   □

Now, we prove that the set of values of the parameter $M \in \mathbb{R}$ for which Green's function is nonpositive on $J \times J$ is connected. The result is, in some sense, a generalization of [13, Corollary 3.1], and it is as follows.

Theorem 1.8.5. *Let $M_1 < \bar{M} < M_2$ be three real constants. Suppose that the nth-order linear two-point boundary value problem* (1.8.1) *has a unique solution for $M = M_j$, $j = 1, 2$, and any $\sigma \in \mathcal{L}^1(J, \mathbb{R})$, and that the corresponding Green's functions satisfy $g_2 \leq g_1 \leq 0$ on $J \times J$. Then problem* (1.8.1) *has a unique solution for $M = \bar{M}$, and the related Green's function $\bar{g}$ satisfies $g_2 \leq \bar{g} \leq g_1 \leq 0$ on $J \times J$.*

Proof. First, note that the uniqueness of solutions of problem (1.8.1) for $M = M_j$, $j = 1, 2$, implies, from Theorem 1.2.10, that the condition of linear independence (1.2.11) or the equivalent one for scalar equations (1.4.4) is fulfilled. As a consequence, to ensure that problem (1.8.1) has a unique solution for $M = \bar{M}$, it suffices to verify that operator $T_n[\bar{M}]$ is surjective on X_U.

Fix $\sigma \geq 0$ in J and let $u_j \leq 0$ on J be the unique solution of problem (1.8.1) with $M = M_j$, $j = 1, 2$. Consider for any $\bar{M} \in (M_1, M_2)$ the problem

$$L_n u(t) = -\bar{M} u(t) + \sigma(t), \ t \in J, \quad U_i(u) = 0, \ i = 1, \ldots, n. \quad (1.8.4)$$

Using that $u_1, u_2 \in X_U$ together with

$$L_n u_2(t) = -M_2 u_2(t) + \sigma(t) \geq -\bar{M} u_2(t) + \sigma(t), \ t \in J, \quad U_i(u_2) = 0, \ i = 1, \ldots, n,$$

and

$$L_n u_1(t) = -M_1 u_1(t) + \sigma(t) \leq -\bar{M} u_1(t) + \sigma(t), \ t \in J, \quad U_i(u_1) = 0, \ i = 1, \ldots, n.$$

We have constructed a pair of well-ordered lower and upper solutions $u_2 \leq u_1$ of problem (1.8.4).

Since $f(t, x) = -\bar{M} x + \sigma(t)$ satisfies that

$$f(t, x) + M_1 x = (M_1 - \bar{M}) x + \sigma(t),$$

is a nonincreasing function in x for any $t \in J$ fixed and operator $T_n[M_1]$ is inverse negative on X_U ($g_1 \leq 0$ in $J \times J$), we can apply Theorem 1.7.2 to ensure that this problem has a solution $\bar{u}$ lying between u_1 and u_2.

In a similar manner we can verify that problem (1.8.4) is solvable for any $\sigma \leq 0$ on J.

As a consequence, from the linearity of operator $T_n[\bar{M}]$ we deduce that it is surjective in X_U. From Theorem 1.2.10 and (1.2.11) we have that there is a unique Green's function $\bar{g}$ related to $T_n[\bar{M}]$. The fact that $g_2 \leq \bar{g} \leq g_1 \leq 0$ on $J \times J$ follows as in the proof of Theorem 1.8.1. $\qquad\qquad\qquad\qquad\qquad\qquad\square$

In the same way as in the previous cases of this section, we can prove the following results for nonnegative Green's functions.

Theorem 1.8.6. *Let M_1, $M_2 \in \mathbb{R}$ and suppose that the nth-order linear two-point boundary value problem (1.8.1) has a unique solution for $M = M_j$, $j = 1, 2$. Let g_j be Green's functions related to the operator $T_n[M_j]$ and suppose that $g_j(t, s) \geq 0$ for all $(t, s) \in J \times J$ and $j = 1, 2$. Then, if $M_1 < M_2$, it is satisfied that $0 \leq g_2 \leq g_1$ on $J \times J$.*

Lemma 1.8.7. *Under the hypotheses of Theorem 1.8.6 the following properties hold:*

(i) For all $t_0 \in (a, b)$ and every $\varepsilon > 0$ there is $t \in J_0 = (t_0 - \varepsilon, t_0 + \varepsilon)$ and an interval $J_t \subset J$ such that $g_1(t, s) > g_2(t, s) \geq 0$ for all $s \in J_t$.
(ii) If $g_2(t_0, s_0) > 0$ for some $(t_0, s_0) \in J \times J$, then there exists $r > 0$ such that $g_1(t, s_0) > g_2(t, s_0) > 0$ for a.e. $t \in (t_0 - r, t_0 + r)$.

Corollary 1.8.8. *Under the hypotheses of Theorem 1.8.6, assuming that $n \leq 2$ and that the coefficient functions in the general nth-order linear operator L_n defined in (1.4.3) are continuous, we have that if $g_2(t_0, s_0) > 0$ for some $(t_0, s_0) \in J \times J$, then there exists $r > 0$ such that $g_1(t, s_0) > g_2(t, s_0) > 0$ for all $t \in (t_0 - r, t_0 + r)$.*

Theorem 1.8.9. *Let $M_1 < \bar{M} < M_2$ be three real constants. Suppose that the nth-order linear two-point boundary value problem (1.8.1) has a unique solution for $M = M_j$, $j = 1, 2$, and any $\sigma \in \mathcal{L}^1(J, \mathbb{R})$, and that the corresponding Green's functions satisfy $0 \leq g_2 \leq g_1$ on $J \times J$. Then problem (1.8.1) has a unique solution for $M = \bar{M}$, and the related Green's function $\bar{g}$ satisfies $0 \leq g_2 \leq \bar{g} \leq g_1$ on $J \times J$.*

Next, we look for a more concise description of the structure of the connected set for which the Green's function related to operator $T_n[M]$ has constant sign. It is obvious that if it is strictly negative or strictly positive in $J \times J$ for some value of M, then this sign property remains valid in a neighborhood of the parameter and the resulting set is an interval. In such a case, it is very important to know if it can be, or not, bounded and, if it is the case, to have some information about the extremes of this interval. On the other hand, it is important to note that there is an important set of problems in which, by its proper definition, Green's function cannot be neither strictly positive nor strictly negative. It is the case, for instance, of the second-order Dirichlet conditions $u(a) = u(b) = 0$, for which the corresponding Green's function must satisfy $g(a, s) = g(b, s) = 0$ for all $s \in (a, b)$. In this case, as it is pointed out in [55], any small perturbation on the parameter M can transform Green's function in a changing sign one.

Thus, we need a more sophisticated analysis, addressed to strongly positive linear operators defined in cones. The results shown here can be found in [55, Chap. 7].

Definition 1.8.10. Let X be a Banach space. A *cone* on X is a closed and convex subset $K \subset X$, such that $\lambda x \in K$ for $x \in K$ and $\lambda \geq 0$ and $K \cap (-K) = \{0\}$.

A cone K defines the partial ordering in X given by $x \preceq y$ if and only if $y - x \in K$. We use the notation $x \prec y$ for $y - x \in K \setminus \{\theta\}$ and $x \not\preceq y$ for $y - x \notin K$, Moreover, $x \ll y$ means $y - x \in \text{int}(K)$.

We say that K is normal if and only if there is a real number $c > 0$ such that if $x, y \in K$ satisfy $0 \preceq x \preceq y$, then $\|x\| \leq c \|y\|$.

Example 1.8.11. If we consider the Banach space of the continuous functions $X = \mathscr{C}(J, \mathbb{R})$, the cone K that defines the usual order in this space is the following one:

$$K = \{u \in \mathscr{C}(J, \mathbb{R}), \quad \text{such that} \quad u(t) \geq 0 \text{ for all } t \in J\}.$$

In this case, $u \prec v$ means $u(t) \leq v(t)$ for all $t \in J$ and there is some $t_0 \in J$ for which $u(t_0) < v(t_0)$.

The interior of K is given as

$$\text{int}(K) = \{u \in \mathscr{C}(J, \mathbb{R}), \quad \text{such that} \quad u(t) > 0 \text{ for all } t \in J\}.$$

As a consequence $u \ll v$ represents $u(t) < v(t)$ for all $t \in J$.

It is immediate to verify that this cone is normal with normal constant $c = 1$.

Let X be a Banach space with an order cone K having a nonempty interior and let $T : X \to X$ be a linear and completely continuous operator. It is very well known (see [30] for details) that the eigenvalues of T, i.e., the set Λ of $\lambda \in \mathbb{C}$ for which there is $x \neq 0$ a solution of $T x = \lambda x$, form a discrete set which may be infinite, finite, or empty. Each eigenvalue is of finite algebraic and geometric multiplicity and 0 is the only limit point if Λ is not finite.

Let $r(T)$ the so-called spectral radius

$$r(T) = \max_{\lambda_i \in \Lambda} |\lambda_i|.$$

We will say that operator T is strongly positive if and only if the following property holds:

$$\theta \prec x \qquad \text{implies} \qquad \theta \ll T x \quad \text{for all } x \in D(T). \tag{1.8.5}$$

It is very well known (see [29] for details) that if T is completely continuous and strongly positive, then $r(T) \in (0, \infty)$.

As we have noted in Example 1.5.3, if $k(t, s) \in \mathscr{L}^2(J \times J, \mathbb{R})$ we know that operator $T_k : \mathscr{L}^2(J, \mathbb{R}) \to \mathscr{L}^2(J, \mathbb{R})$, defined as

$$T_k u(t) = \int_J k(t, s) u(s) ds \tag{1.8.6}$$

is completely continuous.

The next result is a form of the Frobenius-Perron-Jentzsch theorem, given in [30, Theorem 1]

Theorem 1.8.12. *If $k(t,s) \geq 0$ on $J \times J$ and $r(T_k) > 0$ (T_k defined in (1.8.6)), then*

(a) $r(T_k)$ is an eigenvalue of T_k.
(b) There exists a nonnegative eigenfunction corresponding to the eigenvalue $r(T_k)$.

As it is pointed out in [30], there exist kernels $k \geq 0$ on $J \times J$ for which $r(T_k) = 0$. Indeed, consider $k(t,s) = 1$ if $s \in [a,t]$ and $k(t,s) = 0$ elsewhere, which is nonnegative on $J \times J$.

This kernel defines the integral operator

$$T_k \, u(t) = \int_a^t u(s) \, ds.$$

It is obvious that this operator gives us the unique solution of the initial value problem

$$L\,v(t) := v'(t) = u(t), \quad v(a) = 0.$$

In other words, $T_k = L^{-1} : \mathscr{L}^1(J, \mathbb{R}) \to \{v \in \mathscr{AC}(J, \mathbb{R}); \quad v(a) = 0\}$.

Due to the uniqueness of solutions of the initial value problems, we have that operator L has no eigenvalues. In consequence the same occurs with T_k and so $r(T_k) = 0$.

Definition 1.8.13. The operator $T : D(T) \subset X \to X$ is called e-positive if and only if there is an element $e \succ 0$ and, for every $x \in D(T)$, $x \succ 0$, there are positive numbers $\alpha(x)$ and $\beta(x)$ such that

$$\alpha(x)\,e \leq T\,x \leq \beta(x)\,e.$$

Definition 1.8.14. Let X be a real Banach space with an order cone K and let $e \succ 0$. Set

$$X_e = \{x \in X : \quad \text{there is a real } c > 0 \text{ such that} \quad -c\,e \leq x \leq c\,e\},$$

$$\|x\|_e = \inf\{c > 0, \quad -c\,e \leq x \leq c\,e\}$$

and

$$K_e = K \cap X_e.$$

Proposition 1.8.15 ([55, Proposition 7.14]). *If K is normal, then:*

1. The set X_e with the norm $\|\cdot\|_e$ is a Banach space. The embedding $X_e \subset X$ is continuous. If in fact $e \gg 0$, then $X = X_e$ and the norms are equivalent.
2. The set K_e is a normal cone in X and $e \in \mathrm{int}(X_e)$.

3. *If for $x \in X$ there are positive numbers α and β such that $\alpha e \le x \le \beta e$ in X, then $x \gg 0$ in X_e.*
4. *If the linear operator $T : X \to X$ is e-positive and if $T(X) \subset X_e$, then $T : X \to X_e$ is strongly positive.*

Consider the equation

$$Tx = \lambda x, \quad x \succ 0$$

and the correspondent inhomogeneous equation

$$\lambda x - Tx = y, \quad y \succ 0. \tag{1.8.7}$$

Next we enunciate the classical result of Krein and Rutman for the existence and uniqueness of solutions of (1.8.7), depending on the spectral radius of operator T.

Theorem 1.8.16 ([55, Corollary 7.27]). *Let T be a strongly positive operator. For every $y \succ 0$, (1.8.7) has exactly one solution $x \succ 0$ if $\lambda > r(T)$ and no solution $x \succ 0$ if $\lambda \le r(T)$.*

Moreover, given λ, $\mu \in \mathbb{R}$ and $y \succ 0$. If the equation $\lambda x - T x = \mu y$ has a positive solution $x \succ 0$, then $sgn(\mu) = sgn(\lambda - r(T))$.

Remark 1.8.17. For the case $\mu = -1$, we have that if the equation $Tx - \lambda x = y \succ 0$ has a positive solution $x \succ 0$ then $\lambda < r(T)$.

Now, we will study the eigenvalue equation

$$T_n[M] u(t) = \lambda u(t), \ t \in J, \quad U_i(u) = 0, \ i = 0, \ldots, n-1. \tag{1.8.8}$$

If operator $T_n[M]$ is invertible in X_U, the previous differential equation is equivalent to the integral one

$$u(t) = \lambda T_n^{-1}[M] u(t), \ t \in J. \tag{1.8.9}$$

Denoting g_M as the Green's function related to operator $T_n[M]$, we have that

$$T_n^{-1}[M] y(t) = \int_a^b g_M(t,s) \, y(s) \, ds, \quad y \in \mathscr{L}^1(J, \mathbb{R}).$$

Having in mind the case in which Green's function $g_M \le 0$ on $J \times J$ and vanishes at $t = a$ or $t = b$, we assume the following condition:

(N_g) Suppose that there is a continuous function $\phi(t) > 0$ for all $t \in (a, b)$ and k_1, $k_2 \in \mathscr{L}^1(J, \mathbb{R})$, such that $k_1(s) < k_2(s) < 0$ for a.e. $s \in J$, satisfying

$$\phi(t) k_1(s) \le g_M(t,s) \le \phi(t) k_2(s), \quad \text{for a.e. } (t,s) \in J \times J.$$

Example 1.8.18. It is immediate to verify that the Dirichlet problem

$$u''(t) = 0, \ t \in J, \qquad u(a) = u(b) = 0,$$

has $u = 0$ as its unique solution. Moreover, the related Green's function is given by the following expression:

$$g(t,s) = \frac{1}{b-a} \begin{cases} (a-s)(b-t), & \text{if } a \leq s \leq t \leq b, \\[2mm] (a-t)(b-s), & \text{if } a \leq t \leq s \leq b. \end{cases}$$

One can see that condition (N_g) holds for $\phi(t) = \sin(\pi\,(t-a)/(b-a))$, $k_1(s) = \max\{s-b, a-s\}/(\pi(b-a))$ and $k_2(s) = -(\pi/(b-a))^2$.

Remark 1.8.19. We remark that despite condition (N_g) is suitable for Green's functions that vanish at the boundary of $J \times J$, this condition is trivially fulfilled if Green's function is strictly negative. Indeed, let $\gamma < \delta < 0$ be two negative constants such that

$$\gamma \leq g_M(t,s) \leq \delta < 0, \quad \text{for all } (t,s) \in J \times J.$$

It is immediate to verify that condition (N_g) holds by defining $\phi(t) = 1$, $k_1(s) = \gamma$ and $k_2(s) = \delta$, for all $t, \ s \in J$.

Remark 1.8.20. We point out that condition (N_g) is, in some sense, a generalization of the one imposed in [30] in which, in this case for a nonnegative kernel $k(t,s)$, it is assumed that k is continuous and there is ϕ, a strictly positive function on (a,b), for which $k(t,s)/\phi(t)$ is bounded and positive in $J \times J$.

In fact, the choice of the function ϕ in condition (N_g), and even in [30], is given as the eigenfunction associated to the first negative eigenvalue of the operator $T_n[M]$ on X_U.

Remark 1.8.21. We remark that condition (N_g) may not be fulfilled for some nonpositive Green's function. To see this, it suffices to consider the terminal value problem

$$u'(t) + M\,u(t) = 0, \ t \in J, \qquad u(b) = 0.$$

From Lemma 1.5.1, we know that the expression of Green's function related to this problem is given by

$$g_M(t,s) = \begin{cases} 0, & \text{if } a \leq s < t \leq b, \\[2mm] -e^{M(s-t)}, & \text{if } a < t < s \leq b. \end{cases} \tag{1.8.10}$$

It is obvious that such function is nonpositive on $J \times J$ for all $M \in \mathbb{R}$. However, there is no possibility of defining k_2 in condition (N_g).

In next result we prove that condition (N_g) ensures the strongly positive character of operator $-T_n[M]$.

Theorem 1.8.22. *Let X be the Banach space $\mathscr{C}(J, \mathbb{R})$ endowed with the supremum norm $\| \cdot \|_\infty$, and $K = \{x \in X, \ x(t) \geq 0 \ for \ all \ t \in J\}$, the normal cone. Suppose that Green's function g_M related to operator $T_n[M]$ satisfies condition (N_g). Then, denoting $e = \phi \succ 0$, the operator $-T_n^{-1}[M] : X \to X$ is e-positive and strongly positive.*

Proof. Denote $T := -T_n^{-1}[M]$. Let $x \in C(J, \mathbb{R})$ be such that $x \succ 0$ in J. By definition, we have that

$$T x(t) = - \int_a^b g_M(t, s) \, x(s) \, ds \geq -\phi(t) \int_a^b k_2(s) \, x(s) \, ds = \alpha(x) \, \phi(t),$$

with

$$\alpha(x) = - \int_a^b k_2(s) \, x(s) \, ds > 0.$$

Moreover,

$$T x(t) = - \int_a^b g_M(t, s) \, x(s) \, ds \leq -\phi(t) \int_a^b k_1(s) \, x(s) \, ds = \beta(x) \, \phi(t),$$

with

$$\beta(x) = - \int_a^b k_1(s) \, x(s) \, ds > 0.$$

So, we have obtained that T is e-positive in X. Let us see that $T(X) \subset X_e$.

Consider $x \in \mathscr{C}(J, \mathbb{R})$, $x \neq 0$, then denoting as x^+ and x^- its corresponding positive and negative parts on J, we have that

$$T x(t) = - \int_a^b g_M(t, s) \, x(s) \, ds \geq \phi(t) \int_a^b k_1(s) \, x^-(s) \, ds = -\gamma(x) \, \phi(t),$$

and

$$T x(t) = - \int_a^b g_M(t, s) \, x(s) \, ds \leq -\phi(t) \int_a^b k_1(s) \, x^+(s) \, ds = \delta(x) \, \phi(t).$$

As a consequence, choosing $c = \max \{\gamma(x), \delta(x)\} > 0$, we have that $T x \in X_e$. Thus, Proposition 1.8.15 implies that operator T is strongly positive.   □

Now we introduce the set of values in which Green's function is nonpositive on $J \times J$:

$$N_T = \{M \in \mathbb{R}, \quad \text{such that} \quad g_M(t,s) \leq 0 \text{ for all } (t,s) \in J \times J\} \quad (1.8.11)$$

In the following result is described the part of the set N_T on the right of $M \in \mathbb{R}$ for which condition (N_g) holds.

Theorem 1.8.23. *Let $\bar{M} \in \mathbb{R}$ be fixed. If operator $T_n[\bar{M}]$ is invertible in X_U and its related Green's function is nonpositive on $J \times J$ and satisfies condition (N_g), then the following statements hold:*

1. *There exists $\lambda_1 < 0$, the least eigenvalue in absolute value of operator $T_n[\bar{M}]$ in X_U. Moreover, there exists a nontrivial constant sign eigenfunction corresponding to the eigenvalue λ_1.*
2. *Green's function related to operator $T_n[M]$ is nonpositive on $J \times J$ for all $M \in [\bar{M}, \bar{M} - \lambda_1)$.*
3. *Green's function related to operator $T_n[M]$ cannot be nonpositive on $J \times J$ for all $M > \bar{M} - \lambda_1$.*
4. *If there is $M \in \mathbb{R}$ for which Green's function related to operator $T_n[M]$ is nonnegative on $J \times J$, then $M > \bar{M} - \lambda_1$.*

Proof. From Theorem 1.8.22, we are in a position to apply Theorem 1.8.16 to $-T_m^{-1}[\bar{M}]$. In order to do this, we rewrite (1.8.7) to our particular case. First note that from (1.8.8) and (1.8.9), we have that λ is an eigenvalue of operator $T_n[\bar{M}]$ on X_U if and only if $-1/\lambda$ is an eigenvalue of operator $T = -T_n^{-1}[\bar{M}]$.

Since T is completely continuous and strongly positive, we know from [29] that $r(T) > 0$. In particular $\lambda_1 = -1/r(T)$ is an eigenvalue of the differential operator $T_n[\bar{M}]$ in X_U. Moreover, if λ is any eigenvalue of this differential operator, then $|\lambda| \geq 1/r(T)$. So operator $T_n[\bar{M} + \lambda]$ is invertible on X_U for all $|\lambda| < -\lambda_1$, and $\lambda_1 < 0$ is the least eigenvalue in absolute value of operator $T_n[\bar{M}]$ in X_U. The existence of a nontrivial constant sign eigenfunction corresponding to the eigenvalue λ_1 follows from Theorem 1.8.12.

To prove the second assertion, we must study the range of the parameters $\mu > 0$ that satisfy that Green's function related to $T_n[\bar{M} + \mu]$ is nonpositive on $J \times J$. To this end, we get $\sigma \succ 0$ on J and look for the positive values of the parameter μ such that the problem

$$T_n[\bar{M} + \mu] u(t) = \sigma(t), \; t \in J, \quad U_i(u) = 0, \; i = 0, \ldots, n - 1,$$

has a unique solution $u \succ 0$ on J.

This problem is equivalent to the eigenvalue equation

$$T_n[\bar{M}] u(t) = -\mu u(t) + \sigma(t), \; t \in J, \quad U_i(u) = 0, \; i = 0, \ldots, n - 1,$$

or, due to the invertibility of $T_n[\bar{M}]$,

$$u(t) = \mu \left(-T_n^{-1}[\bar{M}]\right) u(t) + T_n^{-1}[\bar{M}] \sigma(t), \ t \in J.$$

So, we arrive at the following expression with $T = -T_n^{-1}[\bar{M}]$:

$$\frac{1}{\mu} \left(-u(t)\right) - T \left(-u(t)\right) = -\frac{1}{\mu} T_n^{-1}[\bar{M}] \sigma(t), \ t \in J. \tag{1.8.12}$$

Now, from condition (N_g), since $\mu > 0$ we deduce that the right-hand side of the previous expression is $\succ 0$. Moreover, the Krein-Rutman theorem 1.8.16 implies that for all $0 < \mu < 1/R(T)$ there is a unique solution $-u \succ 0$ in J of (1.8.12). But this is the same that to say that if $0 < \mu < 1/R(T)$, then Green's function related to operator $T_n[M + \mu]$ is nonpositive on $J \times J$. Since $-1/r(T)$ is an eigenvalue of $T_n[M]$ in X_U, we have that this estimation is optimal.

Assertions 3 and 4 follow as a direct consequence of Theorem 1.8.16 and Remark 1.8.17. $\square$

Example 1.8.24. By direct integration, one can easily check that the eigenvalues of the Dirichlet problem

$$u''(t) = \lambda u(t), \ t \in J, \qquad u(a) = u(b) = 0$$

are given as

$$\lambda_n = -\left(\frac{n\pi}{b - a}\right)^2, \qquad n = 1, 2, \dots$$

Since, as we have seen in Example 1.8.18, Green's function related to operator u'' in $X_U = \{u \in W^{2,1}(J, \mathbb{R}), \ u(a) = u(b) = 0\}$ satisfies condition (N_g), we deduce as a direct consequence of Theorem 1.8.23 the following properties:

1. Green's function related to operator $u'' + M u$ in X_U is nonpositive on $J \times J$ for all $M \in [0, (\pi/(b - a))^2)$.
2. Green's function related to operator $u'' + M u$ in X_U cannot be nonpositive on $J \times J$ for all $M > (\pi/(b - a))^2$.
3. If there is $M \in \mathbb{R}$ for which Green's function related to operator $u'' + M u$ in X_U is nonnegative on $J \times J$, then $M > (\pi/(b - a))^2$.

By making an exhaustive study of Green's function related to this operator, it is proved in [14] that Green's function is, in this case, nonpositive in $J \times J$ for all $M < (\pi/(b - a))^2$ and it changes its sign in $J \times J$ for all $M > (\pi/(b - a))^2$, $M \neq \left(\frac{n\pi}{b-a}\right)^2, n = 1, 2, \dots$

Theorem 1.8.23 gives us the supremum of the values of the real parameter M for which the operator $T_n[M]$ is inverse negative on X_U. In fact, we have that such value is the first eigenvalue of the considered linear differential operator. In particular, it is a singular value of the differential equation. In the previous example we pay attention to the infimum of such interval; in other words, we are interested in knowing what is the entire interval N_T for which, if $M \in N_T$, Green's function related to operator $T_n[M]$ in X_U is nonpositive on $J \times J$.

First, we note that from Theorem 1.8.5, the set N_T is actually an interval. Of course this interval can be empty if condition (N_g) is not fulfilled. This is the case of the initial value problem

$$u'(t) + M\,u(t) = 0, \, t \in J, \quad u(a) = 0,$$

for which, from (1.4.7) and (1.4.10), we have that the expression of its Green's function follows the expression

$$g_M(t,s) = \begin{cases} e^{M(s-t)}, & \text{if } a \le s < t \le b, \\ \\ 0, & \text{if } a < t < s \le b. \end{cases} \tag{1.8.13}$$

It is obvious that it is nonnegative on $J \times J$ for all $M \in \mathbb{R}$.

When condition (N_g) holds, we have that the interval N_T is nonempty. So we are concerned with the infimum of this interval.

First we note that it can be unbounded from below. To see this it is enough to take into account the Dirichlet problem considered in Example 1.8.24, for which $N_T = (-\infty, (\pi/(b-a))^2)$ (see [32] for details).

The question that we are dealing with, in case of N_T is bounded from below, consists of knowing whether the infimum of N_T can or cannot be an eigenvalue of operator $T_n[M]$. The answer is negative and it is given in the following result.

Lemma 1.8.25. *Let $\bar{M} \in \mathbb{R}$ be fixed. Suppose that operator $T_n[\bar{M}]$ is invertible in X_U, its related Green's function is nonpositive on $J \times J$, it satisfies condition (N_g), and the set N_T, defined in (1.8.11), is bounded from below.*

Then $N_T = [\bar{M} - \bar{\mu}, \bar{M} - \lambda_1)$, with $\lambda_1 < 0$ the first eigenvalue of operator $T_n[\bar{M}]$, obtained in Theorem 1.8.23, and $\bar{\mu} \ge 0$ such that $T_n[\bar{M} - \bar{\mu}]$ is invertible in X_U and the related nonpositive Green's function $g_{\bar{M}-\bar{\mu}}$ vanishes at some points of the square $J \times J$.

Proof. From Theorem 1.8.23 we only need to verify that the real number $\bar{c} = \inf\{N_T\} \le \bar{M}$ is a regular point of operator $T_n[M]$ in X_U.

The continuous and decreasing dependence of Green's function with respect to the parameter M, shown in Theorem 1.8.1, tells us that $g_{\bar{M}} \le g_M \le 0$ in $J \times J$ for all $M \in (\bar{c}, \bar{M}]$.

Consider now the following problem:

$$T_n[\bar{c}]\,u(t) = \sigma(t), \, t \in J, \quad U_i(u) = 0, \, i = 1, \dots, n. \tag{1.8.14}$$

It is obvious that for any $M \in (\bar{c}, \bar{M})$, the solutions of the previous problem coincide with the ones of the following one:

$$T_n[M]\,u(t) = (M - \bar{c})\,u(t) + \sigma(t),\ t \in J,\quad U_i(u) = 0,\ i = 1,\dots,n.$$

From Theorem 1.8.5, we know that operator $T_n[M]$ is invertible on X_U for all $M \in (\bar{c}, \bar{M})$. So, the solutions of the previous problem are just the solutions of the integral equation

$$u(t) = (M - \bar{c}) \int_a^b g_M(t,s)\,u(s)\,ds + \int_a^b g_M(t,s)\,\sigma(s)\,ds, \quad t \in J.$$

By denoting

$$F\,u(t) = (M - \bar{c}) \int_a^b g_M(t,s)\,u(s)\,ds, \quad t \in J,$$

this equation is rewritten as

$$(I - F)\,u(t) = \int_a^b g_M(t,s)\,\sigma(s)\,ds, \quad t \in J,$$

with I as the identity operator.

Using that

$$\|F\,u\|_\infty \le |M - \bar{c}| \int_a^b |g_M(t,s)|\,|u(s)|\,ds < |M - \bar{c}| \int_a^b |g_{\bar{M}}(t,s)|\,ds\,\|u\|_\infty,$$

we deduce that for $|M - \bar{c}|$ small enough the norm of F satisfies that $\|F\| < 1$.

This last property implies that $\lambda = 1$ is not an eigenvalue of F. Thus operator $I - F$ is invertible and the uniqueness of solution of problem (1.8.14) is proved, i.e., we have that $\bar{c}$ is a regular point.

The fact that $g_M \le 0$ in $J \times J$ for all $M \in (\bar{c}, \bar{M}]$ implies that $g_{\bar{c}} \le 0$ on $J \times J$, i.e., $\bar{c} \in N_T$.

Finally, we note that if $g_{\bar{c}} < 0$ on $J \times J$, then g_M remains negative in a left neighborhood of $\bar{c}$. But this contradicts the definition of infimum.   □

Next, as a consequence of the previous results and Theorem 1.8.12, we obtain an estimate of the infimum of the interval N_T as follows.

Lemma 1.8.26. *Let $\bar{M} \in \mathbb{R}$ be fixed. Suppose that operator $T_n[\bar{M}]$ is invertible in X_U, its related Green's function is nonpositive on $J \times J$ and satisfies condition (N_g). Let $\lambda_1 < 0$ be the first eigenvalue of operator $T_n[\bar{M}]$ obtained in Theorem 1.8.23; if operator $T_n[\bar{M}]$ has a positive eigenvalue $\lambda_2 > 0$, then*

$$N_T \subset \left[\bar{M} - \frac{\lambda_1 + \lambda_2}{2}, \bar{M} - \lambda_1 \right).$$

Proof. Theorem 1.8.22 tells us that operator $T = -T_n^{-1}[\bar{M}]$ is strongly positive. As a consequence, we have, from Theorem 1.8.16, that its spectral radius $r(T)$

is strictly positive. Thus, Theorem 1.8.12 implies that $r(T)$ is an eigenvalue of operator T. From (1.8.12) and Theorem 1.8.23 we have that $\lambda_1 = -1/r(T) < 0$ is the smallest eigenvalue in absolute value of operator $T_n[\bar{M}]$. In particular $|\lambda_2| \geq |\lambda_1| > 0$.

Assume now, without loss of generality, that λ_2 is the smallest positive eigenvalue of $T_n[\bar{M}]$ on X_U. Suppose now that there is $M_1 \in N_T$ such that $M_1 < \bar{M} - (\lambda_1 + \lambda_2)/2$. It is obvious that $\mu_1 = M_1 - \bar{M} + \lambda_1 < 0$ and $\mu_2 = M_1 - \bar{M} + \lambda_2 > 0$ are two consecutive eigenvalues of operator $T_n[M_1]$. As a consequence, as we have seen in the proof of Theorem 1.8.23, we know that $-1/\mu_1$ and $-1/\mu_2$ are two eigenvalues of the integral operator $-T_n^{-1}[M_1]$.

From the choice of M_1, it is not difficult to verify that $|\mu_1| > |\mu_2|$. Thus, $1/|\mu_1| < 1/|\mu_2|$ and $R(-T_n^{-1}[M_1]) = 1/|\mu_2|$. In such a case, Theorem 1.8.12 ensures that $1/\mu_2 > 0$ is an eigenvalue of $-T_n^{-1}[M_1]$. But this is the same to say that $-\mu_2$ is an eigenvalue of $T_n[M_1]$, i.e., $2(\bar{M} - M_1) - \lambda_2$ is an eigenvalue of $T_n[\bar{M}]$. But, we have that $2(\bar{M} - M_1) - \lambda_2 \in (\lambda_1, \lambda_2)$, in contradiction with the fact that λ_1 and λ_2 are two consecutive eigenvalues of $T_n[\bar{M}]$.   □

An illustration of the previous case is shown in the following example.

Example 1.8.27. If we consider the fourth-order periodic boundary value problem

$$u^{(4)}(t) + M\,u(t) = 0, \quad t \in J, \qquad u^{(i)}(a) = u^{(i)}(b), \; i = 0, 1, 2, 3.$$

It is not difficult to verify that the previous problem has nontrivial solutions if and only if

$$M = -\left(\frac{2n\pi}{b-a}\right)^4, \quad n = 0, 1, \ldots$$

So, if condition (N_g) holds for some negative value of M near to zero, we obtain the a priori estimate

$$N_T \subset \left[-8\left(\frac{\pi}{b-a}\right)^4, 0\right).$$

In [4, Lemma 2.10 and Remark 2.6] it is proved that

$$N_T = \left[-\left(\frac{2\pi\bar{\mu}}{b-a}\right)^4, 0\right),$$

where $\bar{\mu} \approx 0.7528094$ is the unique solution in $(1/2, 1)$ of the equation

$$-\tanh m\pi = \tan m\pi.$$

Moreover, Green's function vanishes at some points of $J \times J$ when $M = -((2\pi\bar{\mu})/(b-a))^4$ and is strictly negative for all the values in the interior of N_T.

Following analogous arguments to the nonpositive case, we can develop a parallel study for nonnegative Green's functions. To this end, we introduce the following condition.

(P_g) Suppose that there is a continuous function $\phi(t) > 0$ for all $t \in (a, b)$ and $k_1, k_2 \in \mathcal{L}^1(J, \mathbb{R})$, such that $0 < k_1(s) < k_2(s)$ for a.e. $s \in J$, satisfying

$$\phi(t) k_1(s) \le g_M(t, s) \le \phi(t) k_2(s), \quad \text{for a.e. } (t, s) \in J \times J.$$

Example 1.8.28. One can verify that Green's function related to the simply supported beam equation studied in [18]

$$u^{(4)}(t) = \sigma(t), \ t \in J, \qquad u(a) = u(b) = u''(a) = u''(b) = 0,$$

follows the expression

$$g(t, s) = \frac{1}{6(b-a)} \begin{cases} (a-s)(b-t)\left(2a(b-s)+t(t-2b)+s^2\right), & \text{if } a \le s \le t \le b, \\ (a-t)(b-s)\left(2a(b-t)-2bs+s^2+t^2\right), & \text{if } a < t < s \le b. \end{cases}$$

In this situation condition (P_g) is fulfilled for

$$\phi(t) = \sin\left(\pi\,(t-a)/(b-a)\right),$$
$$k_1(s) = \min\{b-s, s-a\}\,\pi(b-a)$$

and

$$k_2(s) = (\pi/(b-a))^4.$$

Remark 1.8.29. It is obvious that if Green's function is strictly positive on $J \times J$, then condition (P_g) is trivially fulfilled.

In the same way as in Remark 1.8.20 we have that condition (P_g) is a generalization of the one imposed in [30]. Furthermore the choice of the function ϕ in condition (P_g) is given by the eigenfunction associated to the first positive eigenvalue of operator $T_n[M]$ on X_U.

Remark 1.8.30. Assumption (P_g) is not a generic condition for a nonnegative Green's function. To see this, it suffices to consider the initial value problem

$$u'(t) + M\,u(t) = 0, \ t \in J, \quad u(a) = 0.$$

From (1.8.13) we have that $g_M \ge 0$ on $J \times J$ for all $M \in \mathbb{R}$, but there is no possibility of defining k_1 for which condition (P_g) holds.

Theorem 1.8.31. *Let $\bar{M} \in \mathbb{R}$ be fixed. If operator $T_n[\bar{M}]$ is invertible in X_U and its related Green's function is nonnegative on $J \times J$ and satisfies condition (P_g), then the following statements hold:*

1. *There exists $\lambda_1 > 0$, the least eigenvalue in absolute value of operator $T_n[\bar{M}]$ in X_U. Moreover, there exists a nontrivial constant sign eigenfunction corresponding to the eigenvalue λ_1.*
2. *Green's function related to operator $T_n[M]$ is nonnegative on $J \times J$ for all $M \in (\bar{M} - \lambda_1, \bar{M}]$.*
3. *Green's function related to operator $T_n[M]$ cannot be nonnegative on $J \times J$ for all $M < \bar{M} - \lambda_1$.*
4. *If there is $M \in \mathbb{R}$ for which Green's function related to operator $T_n[M]$ is nonpositive on $J \times J$, then $M < \bar{M} - \lambda_1$.*

Proof. Considering again the Banach space $X = \mathscr{C}(J, \mathbb{R})$ with the norm $\| \cdot \|_\infty$ coupled with the normal cone $K = \{x \in X, \ x(t) \geq 0 \ \text{for all} \ t \in J\}$. Define now the operator $T : X \to X$ as $T x = T_n^{-1}[\bar{M}] x$. Similarly to the nonpositive case we can verify that operator T is e-positive in X, with $e = \phi \succ 0$, and $T(X) \subset X_e$. So it is strongly positive and Theorem 1.8.16 holds.

In this new situation it is immediate to verify that λ is an eigenvalue of operator $T_n[\bar{M}]$ on X_U if and only if $1/\lambda$ is an eigenvalue of operator $T = T_n^{-1}[\bar{M}]$. Thus, $\lambda_1 = 1/r(T) > 0$ is an eigenvalue of the differential operator $T_n[\bar{M}]$ in X_U, and if λ is any other eigenvalue of the differential operator, then $|\lambda| \geq 1/r(T)$. In particular, $T_n[\bar{M} + \lambda]$ is invertible on X_U for all $|\lambda| < 1/r(T)$. As in the nonpositive case the existence of a nontrivial constant sign eigenfunction corresponding to the eigenvalue λ_1 follows from Theorem 1.8.12.

In order to prove the two first assertions, we are dealing with the range of $\mu > 0$ for which $g_{\bar{M}-\mu}$ is nonnegative on $J \times J$. Thus, for any $\sigma \succ 0$ on J given, we study the set of $\mu > 0$ for which the problem

$$T_n[\bar{M} - \mu] u(t) = \sigma(t), \ t \in J, \quad U_i(u) = 0, \ i = 0, \ldots, n-1,$$

has a unique solution $u \succ 0$ on J.

Arguing as in the nonpositive case, we have that this problem is equivalent to the eigenvalue equation

$$u(t) = \mu \, T_n^{-1}[\bar{M}] u(t) + T_n^{-1}[\bar{M}] \sigma(t), \ t \in J,$$

or, which is the same $(T = T_n^{-1}[\bar{M}])$,

$$\frac{1}{\mu} u(t) - T u(t) = \frac{1}{\mu} T \sigma(t), \ t \in J. \tag{1.8.15}$$

Condition (P_g) and $\mu > 0$ imply that $T \sigma \succ 0$. Now the Krein-Rutman theorem ensures that $r(T) > 0$ and that for all $0 < \mu < 1/R(T)$ there is a unique solution $u \succ 0$ in J of (1.8.15). In other words, from Corollary 1.6.12, we have deduced that if $0 < \mu < 1/R(T)$, then Green's function related to operator $T_n[\bar{M} - \mu]$ is nonnegative on $J \times J$. Moreover, since $1/r(T)$ is an eigenvalue of $T_n[\bar{M}]$ in X_U, such estimation is optimal.

The two last assertions are direct consequences of Theorem 1.8.16 and Remark 1.8.17.   □

Example 1.8.32. It is not difficult to verify that the eigenvalues of the simply supported beam equation

$$u^{(4)}(t) = \lambda\, u(t),\ t \in J, \qquad u(a) = u(b) = u''(a) = u''(b) = 0,$$

are given by

$$\lambda_n = \left(\frac{n\,\pi}{b - a}\right)^4, \qquad n = 1, 2, \ldots$$

From Example 1.8.28 we know that Green's function related to operator $u^{(4)}$ on $X_U = \{u \in W^{4,1}(J, \mathbb{R}),\ u(a) = u(b) = u''(a) = u''(b) = 0\}$ satisfies condition (P_g). So, we deduce from Theorem 1.8.31 the following properties:

1. Green's function related to operator $u^{(4)} + M\,u$ in X_U is nonnegative on $J \times J$ for all $M \in (-(\pi/(b - a))^4, 0]$.
2. Green's function related to operator $u^{(4)} + M\,u$ in X_U cannot be nonnegative on $J \times J$ for all $M < -(\pi/(b - a))^4$.
3. If there is $M \in \mathbb{R}$ for which Green's function related to operator $u^{(4)} + M\,u$ in X_U is nonpositive on $J \times J$, then $M < -(\pi/(b - a))^4$.

In [50, Chap. 2, Sect. 4.1.3.] it is proven that Green's function related to this operator is nonnegative on $J \times J$ if and only if

$$-\left(\frac{\pi}{b - a}\right)^4 < M \le 4\left(\frac{k_0}{b - a}\right)^4,$$

with $k_0 \approx 3.9266$ the smallest positive solution of the equation $\tan k = \tanh k$.

Moreover, in [18, Proposition 2.1] it is shown that Green's function is nonpositive on $J \times J$ if and only if

$$-\left(\frac{k_0}{b - a}\right)^4 \le M < -\left(\frac{\pi}{b - a}\right)^4.$$

In this case, Theorem 1.8.31 characterizes the infimum of the values of the real parameter M for which the operator $T_n[M]$ is inverse positive on X_U as the first eigenvalue of the considered linear differential operator. Having in mind the previous remark, we want to describe the entire interval P_T of the real values for which Green's function related to operator $T_n[M]$ in X_U is nonnegative on $J \times J$.

Theorem 1.8.9 tells us that P_T is an interval. It is important to point out that if condition (P_g) is not fulfilled, it can be empty. To see this it suffices to consider the terminal value problem

$$u'(t) + M\,u(t) = 0, \, t \in J, \quad u(b) = 0$$

As it has been noted in Remark 1.8.21, its Green's function follows the expression (1.8.10) and is nonpositive on $J \times J$ for all $M \in \mathbb{R}$.

If condition (P_g) holds, we have, from Theorem 1.8.31, that the interval $P_T \neq \emptyset$. In fact it can be unbounded from above; it is the case of the first-order periodic problem

$$u'(t) + M\,u(t) = 0, \, t \in J, \quad u(a) = u(b),$$

for which, as it is shown in [32], Green's function is strictly positive on $J \times J$ for all $M > 0$.

Introduce now the following set

$$P_T = \{M \in \mathbb{R}, \quad \text{such that} \quad g_M(t, s) \geq 0 \text{ for all } (t, s) \in J \times J\} \qquad (1.8.16)$$

As in the proof of Lemma 1.8.25, one can prove that if P_T is bounded from above, then its supremum is a regular point. The result is the following.

Lemma 1.8.33. *Let $\bar{M} \in \mathbb{R}$ be fixed. Suppose that operator $T_n[\bar{M}]$ is invertible in X_U, its related Green's function is nonnegative on $J \times J$ and satisfies condition (P_g), and the set P_T defined in (1.8.16) is bounded from above.*

Then $P_T = (\bar{M} - \lambda_1, \bar{M} + \bar{\mu}]$, with $\lambda_1 > 0$ the first eigenvalue of operator $T_n[\bar{M}]$ in X_U obtained in Theorem 1.8.31 and $\bar{\mu} \geq 0$ such that $T_n[\bar{M} - \bar{\mu}]$ is invertible in X_U and its nonnegative related Green's function $g_{\bar{M}-\bar{\mu}}$ vanishes at some points of the square $J \times J$.

As in the nonpositive case, from Theorem 1.8.12 we can deduce an estimate of the supremum of the interval P_T.

Lemma 1.8.34. *Let $\bar{M} \in \mathbb{R}$ be fixed. Suppose that operator $T_n[\bar{M}]$ is invertible in X_U, its related Green's function is nonnegative on $J \times J$ and satisfies condition (P_g). Let $\lambda_1 > 0$ be the first eigenvalue of operator $T_n[\bar{M}]$ in X_U defined in Theorem 1.8.31; if operator $T_n[\bar{M}]$ has a negative eigenvalue $\lambda_2 < 0$, then*

$$P_T \subset \left(\bar{M} - \lambda_1, \bar{M} - \frac{\lambda_1 + \lambda_2}{2} \right].$$

Example 1.8.32 shows us that the fourth-order operator $u^{(4)}(t) + M\,u(t)$ defined on the space

$$X_U = \{u \in \mathscr{W}^{4,1}(J, \mathbb{R}), \quad u(a) = u(b) = u''(a) = u''(b) = 0\}$$

has a bounded set P_T.

Since such problem has no negative eigenvalues, Lemma 1.8.34 is not applicable to this situation.

The mentioned Example 1.8.32 presents a problem which parameters M are divided in four intervals distributed from the left to the right as follows: in the unbounded first and fourth intervals Green's function, if it exists, changes its sign on $J \times J$. It is nonpositive on the second (N_T) and nonnegative on the third (P_T). Moreover, the separation between the interval numbers two and three is an eigenvalue of the considered equation.

As we have noted in some of the previous examples, for a given arbitrary differential operator, some of the constant sign intervals (or both in some cases as, for instance, the anti-periodic) can be empty. But our interest now deals with the separation between the constant sign intervals. From Theorems 1.8.23 and 1.8.31 we know that N_T is always on the left of P_T. Our question is if it can be a nondegenerate fifth interval lying between N_T and P_T. The response, as we prove now, is negative.

Theorem 1.8.35. *Let* $\bar{M} \in \mathbb{R}$ *be such that problem* (1.8.1) *has a unique solution for* $M = \bar{M}$ *and the related Green's function* $g_{\bar{M}}$ *satisfies condition* (N_g). *If the interval* P_T, *defined in* (1.8.16), *is nonempty, then* $\sup(N_T) = \inf(P_T)$, *with* N_T *defined in* (1.8.11).

Proof. From Theorem 1.8.23 we know that $[\bar{M}, \bar{M} - \lambda_1) \subset N_T$, with $\lambda_1 < 0$ the first eigenvalue of $T_n[\bar{M}]$. Moreover, Theorem 1.8.12 ensures the existence of a nontrivial and nonnegative function $\psi \in X_U$ such that $T_n[\bar{M} - \lambda_1]\psi = 0$ in J. From (N_g) we have that

$$\frac{1}{\lambda_1}\psi(t) = T_n^{-1}[\bar{M}]\psi(t) = \int_a^b g_M(t,s)\psi(s)\,ds < 0, \quad \text{for all } t \in (a,b),$$

i.e., $\psi > 0$ on $(a.b)$.

Now, suppose that the result is not true, i.e., there is $M_2 \in P_T$ and $M_1 \in (\bar{M} - \lambda_1, M_2)$ for which problem (1.8.1) has a unique solution for $M = M_1$ and g_{M_1} changes its sign in $J \times J$.

As in the proof of Lemma 1.6.5, we can choose $\sigma_1 \in \mathscr{C}^\infty(J, \mathbb{R})$, such that $\sigma_1(t) = 0$ for all $t \in [a, a + \varepsilon] \cup [b - \varepsilon, b]$ for some $\varepsilon > 0$ small enough, and $\sigma_1 \succ 0$ in J, for which the unique solution of problem

$$L_n u(t) = -M_1 u(t) + \sigma_1(t) \equiv f(t, u(t)), \ t \in J, \quad U_i(u) = 0, \ i = 1, \ldots, n,$$

takes some negative values on J.

Since $\psi > 0$ in (a, b) and $\kappa\,\psi$ remains an eigenfunction of operator $T_n[\bar{M} - \lambda_1]$ for any $\kappa \in \mathbb{R}$, we can assume without loss of generality that

$$0 \le \sigma_1(t) \le (M_1 + \lambda_1 - \bar{M})\psi(t) \text{ for all } t \in J.$$

Thus,

$$L_n \, \psi(t) + M_1 \, \psi(t) = (M_1 - \bar{M} + \lambda_1) \, \psi(t) \geq \sigma_1(t), \; t \in J, \quad U_i(\psi) = 0, \; i = 1, \ldots, n.$$

So, $\psi \in X_U$ is a lower solution of this problem.

Moreover, $\beta \equiv 0 \in X_U$ is an upper solution.

Now, the fact that operator $T_n[M_2]$ is inverse positive on X_U coupled with the fact that $f(t, x) + M_2 \, x = (M_2 - M_1) \, x + \sigma_1(t)$ is a nondecreasing function in x allows us, from Theorem 1.7.4, to ensure that the unique solution of the considered problem is lying between 0 and ψ. In particular it is nonnegative in J, in contradiction with the choice of σ_1. $\qquad\square$

In an analogous way we can prove the following result.

Theorem 1.8.36. *Let $\bar{M} \in \mathbb{R}$ be such that problem (1.8.1) has a unique solution for $M = \bar{M}$ and the related Green's function $g_{\bar{M}}$ satisfies condition (P_g). If the interval N_T defined in (1.8.11) is nonempty then $\sup(N_T) = \inf(P_T)$, with P_T defined in (1.8.16).*

To finish this section, we present a result which, in some sense, "closes the circle" between the relationship of the method of lower and upper solutions and the constant sign of Green's functions. On it we obtain some kind of optimal condition to ensure the existence of solutions to nonlinear boundary value problems.

Theorem 1.8.37. *Suppose that there is $\bar{M} \in \mathbb{R}$ such that operator $T_n[\bar{M}]$ is inverse negative on X_U. Then for every $M < \bar{M}$ for which operator $T_n[M]$ is not inverse negative in X_U, we can construct a function f that satisfies condition (H_d) for this M and a pair of well-ordered lower and upper solutions $\alpha \leq \beta$ in J, for which (1.7.1) has no solution lying between α and β.*

Proof. First note that if $T_n[M]$ is not inverse negative in X_U, then the same property holds for all $M' < M$. As a consequence, since $\inf\{N_T\} = \bar{M} - \bar{\mu}$ a regular point, we can assume, without loss of generality, that $T_n[M]$ is invertible in X_U.

Fix $\sigma \geq 0$ in J for which problem (1.8.1) has a solution that attains some positive values on J.

Consider problem

$$L_n \, u(t) = -M \, u(t) + \sigma(t) := f(t, u(t)), \; t \in J, \quad U_i(u) = 0, \; i = 1, \ldots, n.$$

It is obvious that $\beta = 0$ is an upper solution of this problem.

Let α be the unique solution of problem (1.8.1) with $M = \bar{M}$. The inverse negative character of operator $T_n[\bar{M}]$ in X_U ensures that $\alpha \leq 0$ in J. Moreover,

$$L_n \, \alpha(t) = -\bar{M} \, \alpha(t) + \sigma(t) \geq -M \, \alpha(t) + \sigma(t), \; t \in J, \quad U_i(\alpha) = 0, \; i = 1, \ldots, n.$$

So we have a pair of well-ordered lower and upper solutions $\alpha \leq \beta$ of this problem. However, its unique solution is not less than or equals to zero. So it is not in $[\alpha, \beta]$.

We note that in this case $f(t, x) + M_1 x = (M_1 - M) x + \sigma(t)$ is nonincreasing in x if and only if $M_1 \leq M$. But, from Lemma 1.8.25 we know that operator $T_n[M_1]$ cannot be inverse negative on the left of M. $\qquad\qquad\qquad\qquad\qquad\qquad\square$

The dual enunciate is the following:

Theorem 1.8.38. *Suppose that there is $\bar{M} \in \mathbb{R}$ such that operator $T_n[\bar{M}]$ is inverse positive on X_U. Then for every $M > \bar{M}$ for which operator $T_n[M]$ is not inverse positive in X_U, we can construct a function f that satisfies condition (H_i) for this M and a pair of reversed ordered lower and upper solutions $\alpha \geq \beta$ in J, for which (1.7.1) has no solution lying between α and β.*

1.9 Periodic Problems

This section is devoted to the study of maximum and anti-maximum principles for nth-order linear operators coupled with periodic boundary conditions. We will present them in different subsections depending on the order of the consider operator.

Along this section, for any positive integer n, it will be denoted

$$X_n = \{u \in W^{n,1}(J, \mathbb{R}), \ u^{(i)}(a) = u^{(i)}(b), \quad i = 0, \ldots, n-1\} \qquad (1.9.1)$$

and

$$Y_n = \{u \in W^{n,1}(J, \mathbb{R}), \ u^{(i)}(a) = u^{(i)}(b), \ i = 0, \ldots, n-2, \ u^{(n-1)}(a) \geq u^{(n-1)}(b)\}.$$

$$(1.9.2)$$

1.9.1 First-Order Equations

This part is devoted to the study of the values of the real parameter M for which the operator

$$T_1[M] u(t) = u'(t) + M u(t), \quad t \in J,$$

is inverse negative or inverse positive on Y_1.

As we have shown in Example 1.6.8, this problem is equivalent to obtaining the values of $M \in \mathbb{R}$ for which Green's function related to this operator in X_1 is, respectively, nonpositive or nonnegative on $J \times J$.

First, we note that problem

$$T_1[M] u(t) = 0, \ t \in J, \quad u(a) = u(b),$$

has only the trivial solution if and only $M \neq 0$.

Moreover, the adjoint of the operator $u' + M\,u$ is $-u' + M\,u$ and both are defined in X_1.

From Corollary 1.6.16, using the notation of (1.8.11) and (1.8.16), we deduce that $N_T = -P_T$. So, from Theorem 1.8.35, we have that if such intervals are nonempty, then $0 = \inf(P_T) = \sup(N_T)$.

To deduce the expression of this Green's function it suffices to solve the linear equation (1.4.9) for this particular case, i.e.,

$$r'(t) + M\,r(t) = 0, \; t \in J, \quad r(a) - r(b) = 1.$$

By direct integration, it is immediate to verify that its unique solution (for $M \neq 0$) is given by

$$r(t) = \frac{e^{M(a+b-t)}}{e^{bM} - e^{aM}}.$$

Therefore, (1.4.12) tells us that Green's function that we are looking for is

$$g_M(t,s) = \frac{1}{e^{bM} - e^{aM}} \begin{cases} e^{M(b+s-t)}, & \text{if } a \leq s < t \leq b, \\ e^{M(a+s-t)}, & \text{if } a \leq t < s \leq b. \end{cases}$$

It is obvious that $g_M(t,s) < 0$ for all $(t,s) \in (J \times J)\setminus\{(t,t),\, t \in J\}$ if and only if $M < 0$ and $g_M(t,s) > 0$ for all $(t,s) \in (J \times J)\setminus\{(t,t),\, t \in J\}$ if and only if $M > 0$.

In particular we have that condition (N_g) holds for all $M < 0$, (P_g) is fulfilled for all $M > 0$, $N_T = (-\infty, 0)$, and $P_T = (0, \infty)$.

The previous results are exposed in the following lemmas.

Lemma 1.9.1. *Let $u \in W^{1,1}(J, \mathbb{R})$ be such that*

$$u'(t) + M\,u(t) \succ 0, \quad t \in J, \quad u(a) \geq u(b).$$

Then $u(t) > 0$ for all $t \in J$ if and only if $M > 0$.

Lemma 1.9.2. *Let $u \in W^{1,1}(J, \mathbb{R})$ be such that*

$$u'(t) + M\,u(t) \succ 0, \quad t \in J, \quad u(a) \geq u(b).$$

Then $u(t) < 0$ for all $t \in J$ if and only if $M < 0$.

Weaker assertions can be obtained directly from oscillation theory. Indeed, let $M > 0$ and suppose that

$$u'(t) + M\,u(t) \succ 0, \quad t \in J, \quad u(a) \geq u(b).$$

If there is some $t_0 \in J$ for which $u(t_0) < 0$, function u is strictly increasing in $[a, t_0]$. Thus $u(b) < 0$ and the function is strictly increasing in J, which contradicts that $u(a) \geq u(b)$. In consequence $u \succ 0$ in J.

We can deduce $u > 0$ in J whenever $u' + M\, u > 0$ in J.

A similar argument can be done for $M < 0$.

1.9.2 Second-Order Equations

In this part we will make an exhaustive study of Green's function related to the second-order operator $L_{\gamma, M}\, u = u'' + 2\,\gamma\, u' + M\, u$ in the space Y_2 defined in (1.9.2). First, we show that the inverse negative or inverse positive character of two-constant coefficient operators remains valid when both are composed.

Lemma 1.9.3 ([6, Lemma 2.3]). *Let n, $m \in \mathbb{N}$ be two positive integers, and consider the general operators L_n and L_m defined in (1.4.3) and assume that their corresponding coefficients are constants. Suppose that L_n is inverse positive on Y_n and L_m is inverse positive (inverse negative) on Y_m. Then $L_n \circ L_m$ is inverse positive (inverse negative) on Y_{n+m}.*

Proof. Let $u \in Y_{n+m}$. It is clear that $L_m\, u \in Y_n$. Now, since L_n is inverse positive on Y_n, if $L_{m+n}\, u = L_n(L_m\, u)) \geq 0$ on J, we have that $L_m\, u \geq 0$ on J.

Now, using that $u \in Y_m$ together with the inverse positive (inverse negative) character of L_m on Y_m we obtain that $u \geq 0$ ($u \leq 0$) on J.   □

As a direct consequence of the previous result we obtain the following expression for Green's function of the composition of two operators.

Lemma 1.9.4. *Let n, $m \in \mathbb{N}$ be two positive integers, and consider the general operators L_n and L_m defined in (1.4.3) and assume that their corresponding coefficients are constants. Suppose that L_n is invertible in X_n and L_m is invertible on X_m. Denoting as g_n and g_m the corresponding Green's functions, then g_{n+m}, the Green's function related to operator L_{m+n} on X_{m+n}, is given by the following expression:*

$$g_{m+n}(t, s) = \int_a^b g_n(t, \tau)\, g_m(\tau, s)\, d\tau. \tag{1.9.3}$$

Proof. Let $\sigma \in \mathcal{L}^1(J, \mathbb{R})$ and $u \in X_{n+m}$ be the unique solution of the problem

$$L_{m+n}\, u(t) = \sigma(t), \quad t \in J, \qquad u \in X_{m+n}.$$

Since the coefficients are constants we know that $L_m\, u \in X_n$ is the unique solution of the problem

$$L_n\, v(t) = \sigma(t), \quad t \in J, \qquad v \in X_n,$$

that is,

$$L_m u(t) = \int_a^b g_n(t, s) \sigma(s) \, ds \equiv \sigma_1(t), \quad t \in J.$$

Using now that $u \in X_m$, we have that for all $t \in J$ the following equality holds:

$$
\begin{aligned}
u(t) &= \int_a^b g_m(t, s) \sigma_1(s) \, ds \\
&= \int_a^b g_m(t, s) \int_a^b g_n(s, r) \sigma(r) \, dr \, ds \\
&= \int_a^b \left(\int_a^b g_m(t, s) g_n(s, r) \, ds \right) \sigma(r) \, dr,
\end{aligned}
$$

and the proof is concluded.   □

The previous result allows us to deduce the sign of Green's function for some cases of the real parameters γ and M.

It is not difficult to verify that the homogeneous problem

$$L_{\gamma,M} \, u(t) = 0, \quad t \in J, \quad u \in X_2,$$

has nontrivial solutions if and only if one of the two following cases holds:

1. $\gamma \in \mathbb{R}$ and $M = 0$.
2. $\gamma = 0$ and $M = \left(\dfrac{2 k \pi}{b - a} \right)^2$, $\quad k = 0, 1, \ldots$

The adjoint of $L_{\gamma,M}$ is $L_{-\gamma,M}$ and both are defined in X_2. This fact, in contrary to the first-order equation, tells us that the cases $M > 0$ and $M < 0$ are not symmetric. However, we can ensure, from (1.4.14), that for any $M \in \mathbb{R}$ fixed, Green's function related to operator $L_{\gamma,M}$ is the symmetric one to $L_{-\gamma,M}$.

To study the sign of Green's function, we take into account that the characteristic polynomial $\lambda^2 + 2 \gamma \lambda + M$ has the following two roots:

$$\lambda_1 = -\gamma + \sqrt{\gamma^2 - M} \quad \text{and} \quad \lambda_2 = -\gamma - \sqrt{\gamma^2 - M}.$$

So we divide the study in three parts, depending on the qualitative properties of both roots.

Case 1: $M < 0$.

In this situation both roots are real, in particular

$$\lambda^2 + 2 \gamma \lambda + M = (\lambda - \lambda_1) (\lambda - \lambda_2).$$

This expression implies the corresponding one $L_{\gamma,M} = L_{\lambda_1} \circ L_{\lambda_2}$, being $L_{\lambda_i} u = u' - \lambda_i u, i = 1, 2$. Indeed, let $u \in X_2$, then

$$
\begin{aligned}
L_{\lambda_1} \circ L_{\lambda_2} u(t) &= (u'(t) - \lambda_2 u(t))' - \lambda_1 (u'(t) - \lambda_2 u(t)) \\
&= u''(t) - (\lambda_1 + \lambda_2) u'(t) + \lambda_1 \lambda_2 u(t) \\
&= u''(t) + 2\gamma u'(t) + M u(t).
\end{aligned}
$$

Since $M < 0$ we have that $\lambda_1 > 0 > \lambda_2$. From the previous subsection we know that L_{λ_1} is inverse positive on Y_1 and L_{λ_2} is inverse negative on Y_1. As a consequence, Lemma 1.9.3 implies that operator $L_{\gamma,M}$ is inverse negative on Y_2. Moreover, from expression (1.9.3) and the previous subsection, we deduce that the related Green's function is strictly negative on $J \times J$.

Case 2: $0 < M \le \gamma^2$.

In this situation the two roots remain real, but both have the same sign as $-\gamma$. As a consequence the two operators $L_{\lambda_i}, i = 1, 2$, are inverse negative or inverse positive on Y_1 simultaneously. Using Lemma 1.9.3 again, we deduce that operator $L_{\gamma,M}$ is inverse positive on Y_2. Moreover, expression (1.9.3) and the previous subsection tell us that Green's function is strictly positive on $J \times J$.

In the two previous cases, the exact expression of Green's function related to operator $L_{\gamma,M}$ on X_2 is given by (1.4.12), where the unique solution of (1.4.9) is obtained by solving

$$ r''(t) - (\lambda_1 + \lambda_2) r'(t) + \lambda_1 \lambda_2 r(t) = 0, \ t \in J, \quad r(a) = r(b), \ r'(a) = r'(b) + 1. $$

By direct computation we deduce that the expression of function r is given by

$$ r(t) = -\frac{-e^{a\lambda_1 + \lambda_2 t} + e^{a\lambda_2 + \lambda_1 t} + e^{b\lambda_1 + \lambda_2 t} - e^{b\lambda_2 + \lambda_1 t}}{(\lambda_2 - \lambda_1)\left(e^{a\lambda_2} - e^{b\lambda_2}\right)\left(e^{a\lambda_1} - e^{b\lambda_1}\right)}, $$

if $0 < M < \gamma^2$. And it follows the expression

$$ r(t) = \frac{e^{-\gamma t}\left((t-a)e^{-a\gamma} + (b-t)e^{-b\gamma}\right)}{\left(e^{-a\gamma} - e^{-b\gamma}\right)^2} $$

when $0 < M = \gamma^2$.

It is not difficult to verify that $r > 0$ on J if and only if $\lambda_1 > 0 > \lambda_2$ and $r < 0$ on J whenever $\lambda_1 \lambda_2 > 0$.

Case 3: $\gamma^2 < M$.

In this case, the roots of the characteristic polynomial are complex. In consequence the operator $L_{\gamma,M}$ cannot be decomposed into two first-order linear operators.

Now we have that $M = \gamma^2 + \delta^2$ for some $\delta > 0$. To get the values for which Green's function is positive on $J \times J$ (we remark that from Theorem 1.8.35 such function cannot be negative in this case) we must solve the problem

$$r''(t) + 2\gamma\, r'(t) + (\gamma^2 + \delta^2)\, r(t) = 0,\ t \in J, \quad r(a) = r(b),\ r'(a) = r'(b) + 1,$$

which unique solution is

$$r(t) = \frac{e^{\gamma(a+b-t)}\left(e^{a\gamma}\sin(\delta(b-t)) - e^{b\gamma}\sin(\delta(a-t))\right)}{\delta\left(-2e^{\gamma(a+b)}\cos(\delta(a-b)) + e^{2a\gamma} + e^{2b\gamma}\right)}.$$

It is immediate to verify that $r(t) > 0$ for all $t \in J$ if and only if $0 < \delta\,(b-a) < \pi$. Moreover, if $\delta\,(b-a) = \pi$, we have that $r(t) > 0$ for all $t \in (a,b)$ and $r(a) = r(b) = 0$.

We summarize the previous results in the following lemma.

Lemma 1.9.5. *The following properties are satisfied:*

1. *Operator $L_{\gamma,M}$ is inverse negative on Y_2 if and only if $M < 0$*
2. *Operator $L_{\gamma,M}$ is inverse positive on Y_2 if and only if $0 < M \leq \gamma^2 + \left(\frac{\pi}{b-a}\right)^2$.*

The second assertion of the previous result has been proven in [43, Lemma 2.1].

1.9.3 nth-Order Equations

In the two previous sections the optimal estimates on the real parameter M have been obtained for first- and second-order periodic equations. For higher order ones the expressions are more difficult to deal with, and a general expression for any arbitrary natural order n is not known. As we will see in the next sections, each particular case must be studied in detail and, in most of the situations, the information obtained for Green's function in one situation cannot be extrapolated to a different one.

For arbitrary n, only nonoptimal estimations are given. In [52], by making use of the disconjugacy theory [21], the authors obtained the following result.

Lemma 1.9.6. *Let $M > 0\ (M < 0)$ be such that*

$$\mid M \mid < \frac{n^n n!}{\left[\frac{n}{2}\right]^n (b-a)^n (n-1)^{n-1}} \equiv A(n),$$

with $[x]$ the greatest integer smaller than or equal to the real number x.

Then operator $T_n[M]\,u \equiv u^{(n)} + M\,u$ is inverse positive on X_n (inverse negative on X_n).

In [6, Lemma 2.3] it is shown that for all $n \in \mathbb{N}$, $n \geq 2$, it is satisfied that

$$A(2n) < [A(n)]^2.$$

From this fact, by making use of the composition argument of Lemma 1.9.3, the previous result has been improved in [6, Lemma 2.4] for negative values of M as follows.

Lemma 1.9.7. *Let $A(n)$ be defined as in Lemma 1.9.6. It is satisfied that if $M \in [-[A(n)]^2, 0)$, then operator $T_n[M]$ is inverse negative on X_n.*

From Example 1.6.8 we know that the two previous results hold in the space Y_n.

By using the decomposition of the nth-order operator $T_n[M]$ in first- and second-order operators, as an application of Lemma 1.9.3 in [6], better estimations on the values of the parameter M that ensures the inverse positive or inverse negative character of the studied operator are obtained. The result is the following.

Lemma 1.9.8 ([6, Lemma 2.4]). *Operator $T_n[M]$ is inverse positive on Y_n provided that one of the following properties is fulfilled:*

1. $n = 4k, k \in \{1, 2, \ldots\}$ *and* $0 < M \le \left[\dfrac{\pi}{(b-a) \sin\left(\frac{n+2}{2n}\pi\right)}\right]^n.$

2. $n = 2 + 4k, k \in \{1, 2, \ldots\}$ *and* $0 < M \le \left[\dfrac{\pi}{b-a}\right]^n.$

3. n *is odd and* $0 < M \le \left[\dfrac{\pi}{(b-a) \sin\left(\frac{n+1}{2n}\pi\right)}\right]^n.$

Proof. First, we assume that $J = [0, 2\pi]$. From Lemma 1.4.15, to obtain the estimate on the interval $[a, b]$, we must multiply the one attained here by $\left(\frac{2\pi}{b-a}\right)^n$.

Let $m > 0$ be such that $m^n = M$.

First, we suppose that n is even.

In this case, $p(\lambda) = \lambda^n + m^n = 0$ if and only if

$$\lambda = \lambda_l = m\left[\cos\left(\frac{2l+1}{n}\pi\right) \pm i \, \sin\left(\frac{2l+1}{n}\pi\right)\right] \equiv \alpha_l \pm i\,\beta_l,$$

$l = 0, 1, \ldots, \dfrac{n-2}{2}.$

As a consequence we have that

$$\lambda^n + m^n = \prod_{l=0}^{\frac{n-2}{2}} (\lambda^2 - 2\alpha_l \lambda + m^2),$$

and

$$T_n[m^n] \equiv T_0 \circ T_1 \circ \cdots \circ T_{\frac{n-2}{2}}, \tag{1.9.4}$$

with $T_l u = u'' - 2\alpha_l u' + m^2 u.$

If $n = 4k$ for some $k \in \{1, 2, \ldots\}$, then $\beta_l \leq \beta_{\frac{n}{4}} = m \sin\left(\frac{n+2}{2n}\pi\right)$ for all $l \in \{0, 1, \ldots, \frac{n-2}{2}\}$. Thus, using Lemma 1.9.5, if $m \leq \left[2 \sin\left(\frac{n+2}{2n}\pi\right)\right]^{-1}$, then the operator T_l is inverse positive on Y_2 for all $l \in \{0, 1, \ldots, \frac{n-2}{2}\}$. Therefore, Lemma 1.9.3 implies that $T_n[m^n]$ is inverse positive on Y_n.

If $n = 2 + 4k$ for some $k \in \{1, 2, \ldots\}$, then $\beta_l \leq \beta_{\frac{n-2}{4}} = m$ for all $l \in \{0, 1, \ldots, \frac{n-2}{2}\}$ and, as a consequence, T_l is inverse positive on Y_2 when $m \leq \frac{1}{2}$. By (1.9.4) and Lemmas 1.9.5 and 1.9.3 we obtain that $T_n[m^n]$ is inverse positive on Y_n.

Finally we consider the case in which n is odd.

In this case $p(\lambda) = 0$ if and only if $\lambda = -m$ or $\lambda = \lambda_l = \alpha_l \pm i\,\beta_l$, $l = 0, \ldots, \frac{n-3}{2}$.

So

$$\lambda^n + m^n = (\lambda + m) \prod_{l=0}^{\frac{n-3}{2}} (\lambda^2 - 2\alpha_l \lambda + m^2),$$

and

$$T_n[m^n] \equiv T_0 \circ T_1 \circ \cdots \circ T_{\frac{n-3}{2}} \circ S_1,$$

with $S_1 u = u' + m u$.

Now $\beta_l \leq \beta_{\frac{n-1}{4}} = m \sin\left(\frac{n+1}{2n}\pi\right)$ for all $l \in \{0, 1, \ldots, \frac{n-3}{2}\}$.

Thus, if $m \leq \left[2 \sin\left(\frac{n+1}{2n}\pi\right)\right]^{-1}$ arguing as in the even case, we deduce that the operator $T_n[m^n]$ is inverse positive on Y_n. $\square$

For $M < 0$, it is shown in [6, Lemma 2.5] the following analogous result.

Lemma 1.9.9. *Operator $T_n[M]$ is inverse negative on Y_n provided that one of the following properties is fulfilled:*

1. $n = 4k$, $k \in \{1, 2, \ldots\}$ and $-\left[\dfrac{\pi}{b-a}\right]^n \leq M < 0.$

2. $n = 2 + 4k$, $k \in \{1, 2, \ldots\}$ and $-\left[\dfrac{\pi}{(b-a)\sin\left(\frac{n+2}{2n}\pi\right)}\right]^n \leq M < 0.$

3. n is odd and $-\left[\dfrac{\pi}{(b-a)\sin\left(\frac{n+1}{2n}\pi\right)}\right]^n \leq M < 0.$

It is important to point out that, as we will see in the next sections, the previously obtained estimates are not the best possible for all $n \in \mathbb{N}$.

In the two previous results we have obtained lower bounds for the sets N_T and P_T related to operator $T_n[M]$. If we obtain the eigenvalues of operator $T_n[M]$, by using Lemmas 1.8.26 and 1.8.34, depending on the order n of the equation, we could be able to obtain upper bounds for the studied sets. To this end we note that every periodic function $u \in X_n$ can be expressed in a unique form by the following Fourier series:

$$u(t) = a_0 + \sum_{m=0}^{\infty} \left[a_m \cos\left(\frac{2 m \pi (t - a)}{b - a}\right) + b_m \sin\left(\frac{2 m \pi (t - a)}{b - a}\right) \right], \quad t \in J.$$

The convergence in the right-hand side of the equation is uniform in J and the same holds for all the derivatives up to order $n - 1$, in which case the derivative of the Fourier series of a given function coincides with the Fourier series of its derivative. For the nth-order derivative the same property holds, but the convergence is ensured only in $\mathscr{L}^2(J, \mathbb{R})$.

In particular we have that if $u \in X_n$ then

$$u^{(n)}(t) = \sum_{m=0}^{\infty} \left(\tilde{a}_m \cos\left(\frac{2 m \pi (t - a)}{b - a}\right) + \tilde{b}_m \sin\left(\frac{2 m \pi (t - a)}{b - a}\right) \right), \quad t \in J,$$

with

$$\tilde{a}_m = \left(\frac{2 m \pi}{b - a}\right)^n \begin{cases} b_m, & \text{if } n = 4k + 1, \ k = 0, 1, \ldots \\ -a_m, & \text{if } n = 4k + 2, \ k = 0, 1, \ldots \\ -b_m, & \text{if } n = 4k + 3, \ k = 0, 1, \ldots \\ a_m, & \text{if } n = 4k, \quad k = 1, 2, \ldots \end{cases}$$

and

$$\tilde{b}_m = \left(\frac{2 m \pi}{b - a}\right)^n \begin{cases} -a_m, & \text{if } n = 4k + 1, \ k = 0, 1, \ldots \\ -b_m, & \text{if } n = 4k + 2, \ k = 0, 1, \ldots \\ a_m, & \text{if } n = 4k + 3, \ k = 0, 1, \ldots \\ b_m, & \text{if } n = 4k, \quad k = 1, 2, \ldots \end{cases}$$

As a consequence, there is a nontrivial solution of the equation

$$u^{(n)}(t) + M\, u(t) = 0, \ t \in J, \quad u \in X_n,$$

if and only if one of the next situations is fulfilled:

1. $n \in \mathbb{N}$ and $M = 0$.

2. $n = 4k + 2$, for some $k = 0, 1, \ldots$ and $M = \left(\frac{2 m \pi}{b-a}\right)^n$ for some $m = 0, 1, \ldots$

3. $n = 4k$, for some $k = 1, 2, \ldots$ and $M = -\left(\frac{2 m \pi}{b-a}\right)^n$ for some $m = 0, 1, \ldots$

In the first case, we have that $M = 0$ is a simple eigenvalue and any real constant solves the considered equation. For the second and third possibilities we have eigenvalues with multiplicity equal to two. Moreover, the related nontrivial eigenfunctions are given by

$$u(t) = a_m \cos\left(\frac{2m\pi(t-a)}{b-a}\right) + b_m \sin\left(\frac{2m\pi(t-a)}{b-a}\right).$$

From Lemmas 1.9.8 and 1.9.9 we have that N_T and P_T are nonempty sets and that $\sup(N_T) = \inf(P_T) = 0$. Moreover, expression (1.9.3) shows that the related Green's functions do not vanish at any point of $J \times J$ for any value of M in the interior of the intervals obtained in Lemmas 1.9.8 and 1.9.9. In particular conditions (N_g) and (P_g) are fulfilled there. Since the adjoint operator of $u^{(n)} + M u$ is $(-1)^n u^{(n)} + M u$ and it is defined in X_n, we have, from Proposition 1.4.13, that if n is odd, then $N_T = -P_T$.

On the other hand, using Lemmas 1.8.34 and 1.9.8, we deduce that if $n = 4k+2$, for some $k = 0, 1, \ldots$, then

$$\left(0, \left(\frac{\pi}{b-a}\right)^n\right] \subset P_T \subset \left(0, 2^{n-1}\left(\frac{\pi}{b-a}\right)^n\right]. \tag{1.9.5}$$

Analogously, we have, from Lemmas 1.8.26 and 1.9.9, that if $n = 4k$, for some $k = 1, 2, \ldots$, then

$$\left[-\left(\frac{\pi}{b-a}\right)^n, 0\right) \subset N_T \subset \left[-2^{n-1}\left(\frac{\pi}{b-a}\right)^n, 0\right). \tag{1.9.6}$$

1.9.4   Third-Order Equations

Now, we will study the values of the real parameter M for which the third-order differential operator

$$T_3[M]u(t) = u'''(t) + M u(t), \quad t \in J,$$

is inverse negative or inverse positive on the set Y_3.

To this end we must obtain the values of $M \in \mathbb{R}$ for which Green's function related to this operator in X_3 is, respectively, nonpositive or nonnegative on $J \times J$.

As we have seen in the previous section, the only eigenvalue of this problem is $M = 0$. Moreover, the intervals N_T and P_T are nonempty and $0 = \inf(P_T) = \sup(N_T)$. Since the adjoint of the operator $u''' + M u$ is $-u''' + M u$ and both are defined in X_3, we have that $N_T = -P_T$.

Lemma 1.9.8 tells us that

$$\left(0, \left(\frac{2\pi}{\sqrt{3}(b-a)}\right)^3\right] \subset P_T.$$

However, it is not proved that such expression is an equality. To describe the sets N_T and P_T we must study directly the expression of Green's function. In this case (1.4.9) reduces to

$$r'''(t) + M\,r(t) = 0,\ t \in J, \quad r(a) = r(b),\ r'(a) = r'(b),\ r''(a) = r''(b) + 1.$$

Thus, (1.4.12) tells us that the sign of g is the same as the one of r. In the next result it is presented the optimal value for which r is positive on J.

Lemma 1.9.10 ([4, Lemma 2.6]). *Operator $T_3[M]$ is inverse positive on the set Y_3 if and only if*

$$M \in \left(0, \left(\frac{2\pi M_3}{b-a}\right)^3\right].$$

Here $M_3 \approx 0.8832205$ is the unique solution of the equation:

$$\arctan\left(\frac{\sin\sqrt{3}m\pi}{\cos\sqrt{3}m\pi - e^{m\pi}}\right) + \pi = \frac{\sqrt{3}}{3}\log\left(\frac{e^{3m\pi} - e^{m\pi}}{\sqrt{1 + e^{2m\pi} - 2e^{m\pi}\cos\sqrt{3}m\pi}}\right),$$

(1.9.7)

with $\arctan\theta \in \left[-\frac{\pi}{2}, \frac{\pi}{2}\right]$.

Proof. In order to simplify the calculations we will work in the interval $[0, 2\pi]$ and denote $M = m^3$.

By using the algorithm developed in [19] and included in Appendix A of this book, after additional calculations we conclude that for all $m > 0$ it is satisfied that

$$r(t) = p_1(m)e^{m(2\pi-t)} + e^{-\alpha(2\pi-t)}\left\{\left(p_2(m) - \sqrt{3}p_3(m)\right)\cos\beta(2\pi - t)\right.$$

$$\left. + \left(\sqrt{3}p_2(m) + p_3(m)\right)\sin\beta(2\pi - t)\right\},$$

with $\alpha = \dfrac{m}{2}$ and $\beta = \dfrac{\sqrt{3}m}{2}$,

$$p_1(m) = \frac{1}{3m^2(e^{2m\pi} - 1)},$$

$$p_2(m) = \frac{\cos\sqrt{3}m\pi - e^{m\pi}}{3m^2(2\cos\sqrt{3}m\pi - e^{m\pi} - e^{-m\pi})}$$

and

$$p_3(m) = \frac{\sin\sqrt{3}m\pi}{3m^2(2\cos\sqrt{3}m\pi - e^{m\pi} - e^{-m\pi})}.$$

Denoting $s = 2\pi - t$, we have that $r'(s) = 0$ if and only if

$$p_1(m)e^{ms}=e^{-\alpha s}\left\{\left(\sqrt{3}p_2(m)-p_3(m)\right)\sin\beta\, s-\left(p_2(m)+\sqrt{3}p_3(m)\right)\cos\beta\, s\right\}.$$

$$(1.9.8)$$

Along this curve, the function r takes its values on the surface

$$q(s,m) = 2\sqrt{3}e^{-\alpha s}\left\{p_2(m)\sin\beta\, s - p_3(m)\cos\beta\, s\right\}.$$

Moreover, $q(s,m) = 0$ if and only if

$$\tan\beta\, s = \frac{p_3(m)}{p_2(m)} = \frac{\sin\sqrt{3}m\pi}{\cos\sqrt{3}m\pi - e^{m\pi}}.$$

$$(1.9.9)$$

The values of m, for which the function r is equal to zero at its critical points, are given by the intersection between the curve of critical points of r, described in (1.9.8), and the curve of zeros of q, given in (1.9.9). That is,

$$p_1(m)e^{ms} = -\frac{e^{-\alpha s}\left\{p_2^2(m) + p_3^2(m)\right\}\cos\beta\, s}{p_2(m)}.$$

$$(1.9.10)$$

Using expression (1.9.9), we have

$$\cos\beta\, s = \frac{\pm p_2(m)}{\sqrt{p_2^2(m) + p_3^2(m)}}.$$

The first part of the equality (1.9.10) is positive, then $\cos\beta\, s$ is negative. Therefore the equality (1.9.10) holds if and only if

$$L(m,k) = N(m), \quad k \in \mathbb{Z}. \tag{1.9.11}$$

where

$$L(m,k) = \frac{2}{\sqrt{3m}}\left\{\arctan\left(\frac{\sin\sqrt{3}m\pi}{\cos\sqrt{3}m\pi - e^{m\pi}}\right) + k\pi\right\}$$

and

$$N(m) = \frac{2}{3m}\log\left(\frac{e^{3m\pi} - e^{m\pi}}{\sqrt{1 + e^{2m\pi} - 2e^{m\pi}\cos\sqrt{3}m\pi}}\right).$$

Since $L(m,k) \leq \dfrac{\pi(1 + 2k)}{\sqrt{3m}}$, we have that $\cos\beta\, s > 0$ whenever $k \leq 0$. So the equality (1.9.11) is not possible in this situation.

Let us see now that

$$\pi < N(m) < \frac{3\pi}{2} \quad \text{for all } m > 0. \tag{1.9.12}$$

The first inequality is satisfied if and only if

$$e^{\frac{3m\pi}{2}} < \frac{e^{3m\pi} - e^{m\pi}}{\sqrt{1 + e^{2m\pi} - 2e^{m\pi}\cos\sqrt{3}m\pi}},$$

which is equivalent to the fact that

$$g(m) = e^{4m\pi} - e^{3m\pi} + 2e^{2m\pi}\left(\cos\sqrt{3}m\pi - 1\right) - e^{m\pi} + 1 > 0.$$

Since $g(0) = g'(0) = g''(0) = g'''(0) = 0$ and $g^{(4)}(m) > 0$ for all $m > 0$, the first inequality of (1.9.12) is satisfied.

The second inequality is fulfilled if and only if

$$f(m) = e^{\frac{9m\pi}{2}} - e^{4m\pi} + 2e^{\frac{7m\pi}{2}}\cos\sqrt{3}m\pi + e^{\frac{5m\pi}{2}} + 2e^{2m\pi} - 1 > 0.$$

But this is true because of $f(0) = f'(0) = f''(0) = 0$ and $f'''(m) > 0$ for all $m > 0$.

From the fact that $L\left(\frac{1}{\sqrt{3}}, 1\right) = 2\pi$ and $L\left(\frac{2}{\sqrt{3}}, 1\right) = \pi$, we deduce that (1.9.11) has at least one solution in $\left(\frac{1}{\sqrt{3}}, \frac{2}{\sqrt{3}}\right)$ for $k = 1$.

One can verify that $\frac{\sqrt{3}m}{2}L(m, 1)$ has a unique critical point $m_1 \in \left(\frac{1}{\sqrt{3}}, \frac{2}{\sqrt{3}}\right)$, at which a maximum is attained. Moreover, $m_1 < \frac{4}{5}$.

The fact that $\frac{\sqrt{3}\,m}{2}N(m)$ is an increasing function on $\left(\frac{1}{\sqrt{3}}, \frac{2}{\sqrt{3}}\right)$ is a consequence of the following inequality for any m in such interval:

$$2e^{4m\pi} - e^{3m\pi}\left(5\cos\sqrt{3}m\pi + \sqrt{3}\sin\sqrt{3}m\pi\right) + e^{2m\pi} >$$

$$-e^{m\pi}\left(\cos\sqrt{3}m\pi + \sqrt{3}\sin\sqrt{3}m\pi\right) + 1$$

Since $N(4/5) < L(1/\sqrt{3}, 1)$, we obtain that (1.9.11) has a unique solution $M_3 \approx 0.8832205$ on $\left(\frac{1}{\sqrt{3}}, \frac{2}{\sqrt{3}}\right)$.

Analogously we can prove that $N(m) < L(m, 1)$ if $m \leq \dfrac{1}{\sqrt{3}}$ and that $N(m) >$ $L(m, 1)$ if $m \geq \dfrac{2}{\sqrt{3}}$. In consequence (1.9.11) has a unique solution, for $k = 1$, in the whole $\mathbb{R}$.

Due to the fact that function r takes negative values for any m in a small enough right neighborhood of M_3, we have, from Lemma 1.8.33, that such estimate is optimal. $\qquad\qquad\square$

Since $N_T = -P_T$ we have the equivalent statement for the inverse negative operator.

Lemma 1.9.11. *Operator $T_3[M]$ is inverse negative on the set Y_3 if and only if*

$$M \in \left[-\left(\frac{2\pi M_3}{b-a} \right)^3, 0 \right)$$

with M_3 defined in (1.9.7).

The general third-order operator $T_3 u \equiv u''' + A u'' + B u' + C u$ has been studied in different papers [1, 7, 8, 43].

In [8], the case in which the characteristic polynomial $p(\lambda) = \lambda^3 + A\lambda^2 + B\lambda + C$ has three real roots is studied. The comparison principles are presented in bigger sets than Y_3. The results obtained are the following.

Lemma 1.9.12 ([8, Lemma 1.1]). *Let $\lambda_1 \leq \lambda_2 \leq \lambda_3 < 0$ be the real roots of $p(\lambda) = 0$, then operator T_3 is inverse positive on the set*

$$\{u \in W^{3,1}(J, \mathbb{R}), \quad u(a) \geq u(b), \ u'(a) \geq u'(b), \ u''(a) \geq u''(b)\}.$$

Proof. Let $L_i u \equiv u' - \lambda_i u$. From Lemma 1.9.1 we know that such operators are inverse positive on Y_1. Moreover, $T_3 = L_1 \circ L_2 \circ L_3$. The result holds by analogous arguments to the ones used in Lemma 1.9.3. $\qquad\qquad\square$

Analogously, the following comparison results have been proved in different sets.

Lemma 1.9.13 ([8, Lemma 1.2]). *Suppose that the roots of $p(\lambda) = 0$ are $\lambda_1 \leq \lambda_2 < 0 < \lambda_3$. Then operator T_3 is inverse negative on the set*

$$\{u \in W^{3,1}(J, \mathbb{R}), \quad u(a) = u(b), \ u'(a) = u'(b) \ (u'(a) \geq u'(b) \text{ when } \lambda_1 + \lambda_3 \leq 0), \ u''(a) \geq u''(b)\}.$$

Lemma 1.9.14 ([8, Lemma 1.3]). *Suppose that the roots of $p(\lambda) = 0$ are $\lambda_1 < 0 < \lambda_2 \leq \lambda_3$. Then operator T_3 is inverse positive on the set*

$$\{u \in W^{3,1}(J, \mathbb{R}), \quad u(a) = u(b), \ u'(a) = u'(b) \ (u'(a) \leq u'(b) \text{ when } \lambda_1 + \lambda_3 \geq 0), \ u''(a) \geq u''(b)\}.$$

Lemma 1.9.15 ([8, Lemma 1.4]). *Let $0 < \lambda_1 \leq \lambda_2 \leq \lambda_3$ be the real roots of $p(\lambda) = 0$. Then operator T_3 is inverse negative on the set*

$$\{u \in W^{3,1}(J, \mathbb{R}), \quad u(a) \geq u(b), \ u'(a) \leq u'(b), \ u''(a) \geq u''(b)\}.$$

The complex roots of the characteristic polynomial have been studied in [7, 43].

Lemma 1.9.16 ([43]). *Let $\gamma < 0$ and $\alpha \pm i\beta$ be the roots of $p(\lambda) = 0$. Then, if $0 < \beta \leq (\frac{\pi}{b-a})^2$, operator T_3 is inverse positive on Y_3*

Proof. In this case, we have that $T_3 u \equiv (A_1 \circ A_2)u$. Where $A_1 u = u' - \gamma u$ and $A_2 u = u'' - 2\alpha u' + (\alpha^2 + \beta^2) u$.

The proof holds from Lemmas 1.9.3 and 1.9.5. □

We note that in [43] the proof is given in X_3.

In [7, Theorem 4.2] the exact values of the roots of the characteristic polynomial of operator T_3, which ensures the positiveness of Green's function, have been obtained.

1.9.5   Fourth-Order Equations

In this subsection we deal with the values of the parameter M for which the fourth-order differential operator

$$T_4[M] u(t) = u^{(4)}(t) + M u(t), \quad t \in J,$$

is inverse negative or inverse positive on Y_4. As we have notice along this section, such problem is equivalent to find the set of $M \in \mathbb{R}$ for which Green's function related to this operator in X_4 has constant sign.

In Sect. 1.9.3 we have proved that the homogeneous problem

$$T_4[M] u(t) = 0, \ t \in J, \quad u \in X_4,$$

has nontrivial solutions if and only if

$$M = -\left(\frac{2 k \pi}{b - a}\right)^4, \quad k = 0, 1\ 2\ \ldots$$

Moreover, this is a self-adjoint operator. Thus, as in the second-order case, the structures of N_T and P_T are completely independent.

Lemma 1.9.9 tells us that operator $T_4[M]$ is inverse negative on Y_4 for all $M \in [-\left(\frac{\pi}{b-a}\right)^4, 0)$ and that its related Green's function is strictly negative for all $M \in (-\left(\frac{\pi}{b-a}\right)^4, 0)$. So, for such a values, condition (N_g) is fulfilled.

From (1.9.6) we know that

$$\left[-\left(\frac{\pi}{b-a}\right)^4, 0\right) \subset N_T \subset \left[-8\left(\frac{\pi}{b-a}\right)^4, 0\right).$$

Furthermore, Lemma 1.9.8 implies that

$$\left(0, \ 4 \left(\frac{\pi}{b-a}\right)^4\right] \subset P_T.$$

Moreover, the related Green's function is strictly positive for M in the interior of the previous interval and, as a consequence, condition (P_g) holds.

In this case, (1.4.9) is as follows

$$r^{(4)}(t)+M \, r(t)=0, \ t \in J, \quad r^{(i)}(a)=r^{(i)}(b), \ i=0,1,2, \quad r'''(a)=r'''(b)+1.$$

Next, we characterize the set N_T.

Lemma 1.9.17 ([4, Lemma 2.10]). *Operator $T_4[M]$ is inverse negative on the set Y_4 if and only if*

$$M \in \left[-\left(\frac{2\pi M_4}{b-a}\right)^4, 0\right).$$

Here $M_4 \approx 0.7528094$ is the unique solution in $\left(\frac{1}{2}, 1\right)$ of the equation:

$$-\tanh m\pi = \tan m\pi$$

Proof. We consider the interval $[0, 2\pi]$ and denote $M = -m^4$ for $m > 0$. For the general interval $[a, b]$ we use Lemma 1.4.15.

In this case the function r is well defined for all $m \notin \mathbb{N}$ and, in such a case, it is given by the expression

$$r(t) = \frac{R(t)}{4m^3 \, (1 - e^{2m\pi}) \, (1 - e^{-2m\pi}) \, (1 - \cos 2m\pi)},$$

with $R(t) = f(t) + g(t)$. Where the functions f and g are given by

$$f(t) = 4 \, (\sin m(2\pi - t) + \sin m \, t) \sinh^2 m \, \pi$$

and

$$g(t) = 2(\sinh m(2\pi - t) + \sinh m \, t)(1 - \cos 2m\pi).$$

It is clear that $g \geq 0$ in $[0, 2\pi]$ for all $m > 0$ and that

$$f(t) \geq 0 \text{ for all } t \in [0, 2\pi] \text{ if and only if } m \in \left(0, \frac{1}{2}\right].$$

Since that $f(2\pi - t) = f(t)$ it suffices to study the function f in the interval $[0, \pi]$. If $m \in \left(\dfrac{1}{2}, 1\right)$, the unique root of f in the interval $[0, \pi]$ is $\dfrac{(2m - 1)\pi}{2m}$. In consequence, using that $f(\pi) = 8 \sin m\pi \sinh^2 m\pi$, we obtain that $f \geq 0$ $(R \geq 0)$ on $\left[\dfrac{(2m - 1)\pi}{2m}, \pi\right]$ when $m \in \left(\dfrac{1}{2}, 1\right)$.

It is easy to verify that $R'' \geq 0$ in the interval $\left[0, \dfrac{(2m - 1)\pi}{2m}\right)$. Since $R'(0) = 0$ the function R attains its minimum at $t = 0$. In consequence the greatest value of m for which the function R is positive in $[0, 2\pi]$ will be the smallest positive root of the expression

$$R(0) = 2(1 - \cos 2m\pi) \sinh 2m\pi + 4 \sinh^2 m\pi \sin 2m\pi.$$

It is obvious that $R(0) = 0$ if and only if either $m \in \mathbb{N}$ or

$$-\tanh m\pi = \tan m\pi.$$

Since $r(0) = r(2\pi) < 0$ for m bigger and close enough to M_4, Lemma 1.8.33 ensures that such estimate is optimal.   □

The case of $M > 0$ has been studied in [14]. On that paper, the set P_T is described as follows.

Lemma 1.9.18 ([13, Theorem 4.1]). *The operator $T_4[M]$ is inverse positive on Y_4 if and only if*

$$M \in \left(0, 4 \left(\frac{2\pi M_4}{b - a}\right)^4\right],$$

with M_4 defined as in Lemma 1.9.17.

Proof. Let $M = m^4$, with $m > 0$, suppose that $[a, b] = [0, 2\pi]$ and denote by $\alpha = m/\sqrt{2}$.

In this case the expression of the function r is given by

$$r(t) = \frac{e^{2\pi\alpha}}{\sqrt{2} m^3 F(\alpha)} \{\sin \alpha(2\pi - t) \cosh \alpha t + \cos \alpha(2\pi - t) \sinh \alpha t$$

$$+ \cos \alpha t \sinh \alpha(2\pi - t) + \sin \alpha t \cosh \alpha(2\pi - t)\},$$

with $F(\alpha) = \left(e^{2\pi\alpha} \sin 2\pi\alpha\right)^2 + \left(1 - e^{2\pi\alpha} \cos 2\pi\alpha\right)^2$.

As we have seen at the beginning of this subsection, operator $T_4[m^4]$ is inverse positive on Y_4 for all $\alpha \in (0, 1/2]$, so we have that $r \geq 0$ on J for all α in such interval and we only must study the sign of the function r for $\alpha > 1/2$.

First, we take into account that the self-adjoint character of operator $T_4[M]$ implies that $r(t) = r(2\pi - t)$. Thus, we reduce our attention to the interval $[0, \pi]$.

Notice that $r(0) = r(2\pi) > 0$ and $r'(0) = r'(\pi) = 0$ for all $\alpha > 0$. Fix $\alpha \in [3/4, 1)$. Let us prove that $r'(t) < 0$ for all $t \in (0, \pi)$.

First, we consider $t \in J_1 = (0, 2\pi - \pi/\alpha]$ (note that $2\pi - \pi/\alpha < \pi$ if and only if $\alpha < 1$). In this case, $\alpha t \in (0, \pi)$ and $\alpha(2\pi - t) \in [\pi, 2\pi)$, consequently $r' < 0$ in J_1.

Now, let $t \in J_2 = (2\pi - \pi/\alpha, \pi) \subset (\pi/2\alpha, \pi/\alpha)$. Therefore $\alpha t, \alpha(2\pi - t) \in (\pi/2, \pi)$. Thus, $r''(t) > 0$ for all $t \in J_2$, which implies that r' is an increasing function in J_2. Since $r'(\pi) = 0$, we conclude that $r' < 0$ in J_2.

So, it is proven that $r'(t) < 0$ for all $t \in (0, \pi)$ and $\alpha \in [3/4, 1)$ and we deduce that $r(t) > r(\pi)$ for all $t \in [0, 2\pi]$ and $\alpha \in [3/4, 1)$.

Since for $\alpha = 3/4$ we have that $r(\pi) > 0$, we deduce that $T_4[M]$ is inverse positive on Y_4 for all $\alpha < 3/4$.

Now, $r(\pi) = 0$ if and only if equation

$$\tan \pi \frac{\sqrt{2}}{2} m = - \tanh \pi \frac{\sqrt{2}}{2} m.$$

It is obvious that the smallest positive solution of such expression is given by $m_4 = \sqrt{2}\, M_4$. The result holds from the fact that such root is simple and by Lemma 1.8.33.   □

1.9.6 Sixth-Order Equations

This part is devoted to the study of the negative values of M for which the sixth-order differential operator

$$T_6[M]\, u(t) = u^{(6)}(t) + M\, u(t), \quad t \in J,$$

is inverse negative on Y_6.

In Sect. 1.9.3 we have proved that the eigenvalues of this operator in X_6 are given by

$$M = \left(\frac{2k\pi}{b-a} \right)^6, \quad k = 0, 1, 2, \ldots$$

In particular, we have no a priori upper bound for N_T. Moreover, since it is a self-adjoint operator, the structure of P_T does not give us any information about N_T.

Now, from the fact that for any $M < 0$ it is fulfilled that $T_6[M] = T_3[\sqrt{-M}] \circ T_3[-\sqrt{-M}]$, we conclude, by Lemmas 1.9.3 and 1.9.10 , that

$$\left[-\left(\frac{2\pi M_3}{b-a} \right)^6, 0 \right) \subset N_T,$$

with M_3 defined as in Lemma 1.9.10.

Moreover, the related Green's function is strictly positive for M in the interior of the previous interval and, as a consequence, condition (P_g) holds.

Now, denoting $M = -m^6$ for some $m > 0$, (1.4.9) is as follows:

$$r^{(6)}(t) - m^6 r(t) = 0, \ t \in J, \quad r^{(i)}(a) = r^{(i)}(b), \ i = 0, \ldots, 4, \quad r^{(5)}(a) = r^{(5)}(b) + 1.$$

Next, the set N_T is characterized.

Lemma 1.9.19 ([13, Theorem 4.2]). *Operator $T_6[M]$ is inverse negative on Y_6 if and only if*

$$M \in \left[-\left(\frac{2\pi M_6}{b-a} \right)^6, 0 \right),$$

where $M_6 \approx 1.010105$ is the unique solution in $\left(\sqrt{3}/2, 2/\sqrt{3} \right)$ of the equation

$$\sinh m\pi \left(2 \sinh m\,\pi/2 \cos \sqrt{3}\,m\pi/2 + \sqrt{3} \cosh m\,\pi/2 \sin \sqrt{3}\,m\pi/2 \right)$$

$$= \cos \sqrt{3}\,m\pi - \cosh m\,\pi.$$

Proof. Consider the interval $[0, 2\pi]$. From the proof of Lemma 1.9.10, by numerical methods, it is not difficult to verify that $M_3 > \sqrt{3}/2$. As a consequence, we study the sign of Green's function related to $T_6^{-1}[-m^6]$ for $m > \sqrt{3}/2$.

In this case, the unique solution of (1.4.9) is given by

$$r(t) = -\frac{w(t) + w(2\pi - t)}{2m^3}. \tag{1.9.13}$$

Here w is the unique solution of the third-order linear problem

$$u'''(t) + m^3 u(t) = 0, \quad u^{(i)}(0) = u^{(i)}(2\pi), \ i = 0, 1, \quad u''(0) - u''(2\pi) = 1, \tag{1.9.14}$$

that has been obtained in the proof of Lemma 1.9.10.

So, the expression of the function r is given by

$$\begin{aligned}
r(t) = \ &\frac{-1}{6m^5 F(m)} \Big\{ \left(-\cos \sqrt{3}\,m\pi - \sqrt{3} \sin \sqrt{3}\,m\pi + e^{-m\pi} \right) e^{at} \cos bt \\
&+ \left(\sqrt{3} \cos \sqrt{3}\,m\pi - \sin \sqrt{3}\,m\pi - \sqrt{3}\,e^{-m\pi} \right) e^{at} \sin bt \\
&+ \left(-e^{m\pi} + \cos \sqrt{3}\,m\pi - \sqrt{3} \sin \sqrt{3}\,m\pi \right) e^{-at} \cos bt \\
&+ \left(-\sqrt{3}\,e^{m\pi} + \sin \sqrt{3}\,m\pi + \sqrt{3} \cos \sqrt{3}\,m\pi \right) e^{-at} \sin bt \Big\} \\
&- \frac{1}{6m^5 G(m)} \left\{ e^{-mt} + e^{-m(2\pi - t)} \right\},
\end{aligned}$$

with $F(m) = 2 \cos \sqrt{3}\, m\pi - e^{m\pi} - e^{-m\pi} < 0$ for all $m > 0$, $G(m) = 1 - e^{-2m\pi} > 0$ for all $m > 0$, $a = m/2$ and $b = \sqrt{3}\, a$.

Using expression (1.9.13), we see that $r(t) = r(2\pi - t)$ (we know this property because $T_6[M]$ is self-adjoint in X_6). Therefore, we only study the function r in the interval $[0, \pi]$. Furthermore, using (1.9.13) and (1.9.14), we deduce that $r'(0) = r'(\pi) = 0$ (this property also holds from Corollary 1.4.12).

Denote $I = (\sqrt{3}/2, 2/\sqrt{3})$, so

$$r''(0) = -(2 \cos \sqrt{3}m\pi \cosh m\pi - 2 \cosh^2 m\pi + 4 \sinh^2 m\pi)/(3m^3 e^{m\pi} F(m) G(m)).$$

Since $\cos \sqrt{3}m\pi \geq 0$ for all $m \in I$ and $-2 \cosh^2 m\pi + 4 \sinh^2 m\pi > 0$ for all $m \in I$, we conclude that $r''(0) > 0$ for all $m \in I$.

If $t \in [t_0, \pi)$, with $t_0 = \pi/(\sqrt{3}m)$ and $m \in I$, we have that $bt \in [\pi/2, \pi)$ and $\sqrt{3}m\pi - bt \in (3\pi/4, 3\pi/2)$. In consequence, using that

$$r''(t) = \frac{-2}{3\,m^3 F(m)} \left\{ \sinh at \cos (\sqrt{3}\, m\pi - bt) + \sinh (m\pi - at) \cos bt \right\}$$

$$- \frac{1}{6\,m^3 G(m)} \left\{ e^{-mt} + e^{-m(2\pi - t)} \right\},$$

we obtain that $r'' < 0$ in $[t_0, \pi]$.

Since $r'(\pi) = 0$, we deduce that $r' > 0$ in $[t_0, \pi)$.

On the other hand, by direct differentiation we obtain that

$$r^{(4)}(0) = \frac{x_1(m) - x_2(m)}{3\, m\, F(m)\, G(m)},$$

with $x_1(m) = -\left(1 + e^{-2m\pi}\right) \cos \sqrt{3}\, m\pi - \sqrt{3} \left(1 - e^{-2m\pi}\right) \sin \sqrt{3}\, m\pi$ and $x_2(m) = -e^{m\pi} - e^{-3m\pi}$.

Using that $x_1^{(i)}(0) = x_2^{(i)}(0)$, $i = 0, 1, 2, 3$, $x_1^{(4)}(m) > x_2^{(4)}(m)$ for all $m \in \left(0, 1/\sqrt{3}\right)$ and $x_2\left(1/\sqrt{3}\right) < \min \{x_1(m); m > 1/\sqrt{3}\}$, we deduce that $r^{(4)}(0) < 0$ for all $m > 0$.

Let $m \in I$ be fixed. Next, we will prove that $r' > 0$ on $I_0 = (0, t_0)$. To this end, let $t \in I_0$. In this case $bt \in (0, \pi/2)$. Thus, since

$$r^{(5)}(t) = \frac{1}{6} \left[\frac{f(t)}{F(m)} + \frac{g(t)}{G(m)} \right],$$

with

$$f(t) = 4 \left[\left\{ \cosh at \cos \sqrt{3}\, m\pi - \cosh a(2\pi - t) \right\} \cos bt \right.$$

$$\left. + \cosh at \sin \sqrt{3}\, m\pi \sin bt \right]$$

and

$$g(t) = e^{-mt} - e^{-m(2\pi - t)},$$

we deduce, by using that $\cosh at \cos \sqrt{3}\,m\pi - \cosh a(2\pi - t) < 0$, that $r^{(5)}(t) > 0$.

Thus, if $r^{(4)}(t_0) \leq 0$, then $r^{(4)} < 0$ in I_0. Therefore, r''' is a decreasing function in I_0. Using (1.9.13) we know that $r'''(0) = 0$; therefore, $r''' < 0$ in I_0.

On the other hand, we have proven in the previous case that $r''(t_0) < 0$. Since $r''(0) > 0$, there exists a unique $t_1 \in I_0$ such that $r''(t_1) = 0$, $r'' > 0$ in $[0, t_1)$ and $r'' < 0$ in (t_1, t_0). Now, due to the fact that $r'(0) = 0$, we know that $r' > 0$ in $(0, t_1)$. Furthermore, since $r'(t_0) > 0$ we deduce that $r' > 0$ in $[t_1, t_0]$.

Thus, we have proven that if $r^{(4)}(t_0) \leq 0$, then $r'(t) > 0$ for all $t \in (0, \pi)$.

Now, let us consider the other case, that is, $r^{(4)}(t_0) > 0$.

In this case, since $r^{(5)} > 0$ in I_0 and $r^{(4)}(0) < 0$, there exists a unique $t_2 \in I_0$ such that $r^{(4)}(t_2) = 0$, $r^{(4)} < 0$ in $(0, t_2)$ and $r^{(4)} > 0$ in (t_2, t_0). From the fact that $r'''(0) = 0$ we deduce that $r''' < 0$ in $(0, t_2]$. Moreover, r''' is an increasing function in $(t_2, t_0]$. Thus, if $r'''(t_0) \leq 0$, we deduce that $r''' < 0$ in $(0, t_0)$. Reasoning as in the previous case, we deduce that $r'(t) > 0$ for all $t \in (0, t_0]$.

Now, let us assume that $r'''(t_0) > 0$.

Since $r'''(t_2) < 0$ there exists a unique $t_3 \in (t_2, t_0)$ such that $r'''(t_3) = 0$, $r''' < 0$ in $(0, t_3)$ and $r''' > 0$ in $(t_3, t_0]$. Due to the fact that $r''(t_0) < 0$, we know that $r'' < 0$ in $[t_3, t_0]$. Taking into account that $r'(t_0) > 0$, we conclude that $r' > 0$ in $[t_3, t_0]$ and, in consequence, $r' > 0$ in $[t_3, \pi]$.

To prove that $r' > 0$ in $(0, t_3)$, using $r''(0) > 0$ and $r''(t_3) < 0$, we deduce that there exists a unique $t_4 \in (0, t_3)$ such that $r''(t_4) = 0$, $r'' > 0$ in $[0, t_4)$ and $r'' < 0$ in $(t_4, t_3]$. Thus, since $r'(0) = 0$ and $r'(t_3) > 0$, we obtain that $r' > 0$ in $(0, t_3)$.

So, it is proven that if $r^{(4)}(t_0) > 0$, then $r' > 0$ in $(0, \pi)$.

Thus, we conclude that $r'(t) > 0$ for all $t \in (0, \pi)$ and $m \in I$, and hence, $t = \pi$ is the unique maximum of r in $[0, 2\pi]$ for all $m \in I$. Then, we can affirm that for all $m \in I$, $R(t) \leq 0$ if and only if $R(\pi) \leq 0$.

Now, since

$$R(\pi) = -\frac{1}{m^3}\,w(\pi),$$

we deduce that $r(\pi) = 0$ if and only if $w(\pi) = 0$. This expression is given in the statement of the Lemma.

It is not difficult to verify that $r(\pi)$ has a unique simple root in I. Thus, Lemma 1.8.25 ensures that M_6 is the greatest value of the constant $m > 0$ for which the operator $T_6[M]$ is inverse negative on Y_6.   □

1.10 Separated Problems

This section is devoted to the study of different linear problems coupled with separated boundary conditions. That is, we are considering two-point boundary conditions defined as (1.4.2), such that for every $i \in \{1, \ldots, n\}$ the functional U_i takes its values only at a or at b. In other words, for any i fixed, either $\alpha_j^i = 0$ or $\beta_j^i = 0$ for every $j = 0, \ldots, n - 1$.

In particular, we will consider the general second-order separated boundary conditions and the fourth-order clamped and simply supported beam equations.

1.10.1 Second-Order Separated Boundary Conditions

In this part we deal with the study of the operator $T_2[M] u = u'' + M u$ defined on the set

$$Z = \left\{ u \in W^{2,1}(J, \mathbb{R}); \qquad B_0 u(a) \leq 0, \qquad B_1 u(b) \leq 0 \right\},$$

with $B_0 u(a) = p_0 u(a) - q_0 u'(a)$ and $B_1 u(b) = p_1 u(b) + q_1 u'(b)$.

Here $M \in \mathbb{R}$, $p_0, p_1, q_0, q_1 \geq 0$, $p_0 + q_0 > 0$, $p_1 + q_1 > 0$.

The case of negative values of M has been considered by different authors. In [15, 32] one can find the following result.

Lemma 1.10.1 ([15, Lemma 2.1]). *The operator $T_2[M]$ is inverse negative on Z for all $M < 0$.*

Proof. First, suppose that $u(t) > 0$ for every $t \in J$.

If either $q_0 = 0$ or $q_1 = 0$, we obtain that $u(a) \leq 0$ or $u(b) \leq 0$ respectively, which is not possible. Thus, we have that $q_0 > 0$ and $q_1 > 0$ and we get $u'(a) \geq 0 \geq u'(b)$, but this contradicts the fact that $u'' \geq -M u > 0$ a.e. $t \in J$.

Hence, there exists $t_1 \in J$ with $u(t_1) \leq 0$.

Let $t_0 \in J$ with $u(t_0) = \max_{t \in J} \{u(t)\} > 0$.

If $t_0 \in (a, b)$, then $u'(t_0) = 0$. But, in this case $u'' \geq -M u$ is positive a.e. in a neighborhood of t_0, which is not possible since u attains its maximum at t_0.

Suppose that $t_0 = a$, thus $u'(a) \leq 0$. If $p_0 > 0$, since $p_0 u(a) \leq q_0 u'(a)$, we get that $u(a) \leq 0$, in contradiction with $u(a) > 0$.

If $p_0 = 0$ we have that $u'(a) = 0$. Let $t_2 \in (a, b)$ be such that $u(t_2) = 0$ and $u(t) > 0$ for all $t \in [a, t_2)$; therefore, $u'' > 0$ a.e. $t \in [a, t_2)$. Now, $u'(a) = 0$ implies that u is increasing in $[a, t_2)$. Thus $0 < u(a) < u(t_2) = 0$ and we arrive to a contradiction.

In an analogous way we conclude that $t_0 \neq b$. In consequence $u \leq 0$ in J. □

In [14] the positive values of the parameter M have been considered. To this end, the following problem that we will denote as $(P_{L_0, L_1}[M])$ has been studied :

$$T_2[M] u(t) = \sigma(t) \text{ a. e. } t \in J, \qquad B_0 u(a) = L_0, \quad B_1 u(b) = L_1.$$

As we have noted in Remark 1.2.14, if there is a unique solution u, then it is given by the expression

$$u(t) = \int_a^b g_M(t,s)\,\sigma(s)\,ds + h_M(t, L_0, L_1), \qquad (1.10.1)$$

with h_M the unique solution of $(P_{L_0,L_1}[M])$ for $\sigma \equiv 0$.

First, the case $M = 0$ has been considered.

Lemma 1.10.2 ([14, Lemma 3.3]). *The operator $L[0]$ is inverse negative on Z if and only if $p_0 > 0$ or $p_1 > 0$.*

Proof. Consider $J = [0, 2\pi]$. In this case Green's function is given by

$$g(t,s) = -\frac{1}{k} \begin{cases} g_1(t,s) := (p_1(2\pi - t) + q_1)\,(s\,p_0 + q_0)\,, & 0 \le s \le t \le 2\pi \\ g_2(t,s) := g_1(s,t), & 0 \le t \le s \le 2\pi, \end{cases}$$

where $k = 2\pi p_0 p_1 + p_0 q_1 + p_1 q_0$.

The function $h_0(t, L_0, L_1)$ is defined by the following expression:

$$h_0(t, L_0, L_1) = \frac{L_0\,(p_1(2\pi - t) + q_1) + L_1\,(q_0 + t\,p_0)}{k}.$$

It is obvious that problem $(P_{L_0,L_1}[0])$ has a unique solution if and only if $k \ne 0$. Moreover, $k = 0$ if and only if $p_0 = p_1 = 0$.

Obviously, $g(t,s) \le 0$ for all $(t,s) \in J \times J$.

On the other hand, since $L_0, L_1 \le 0$ we have that $h_0(t) \le 0$ for all $t \in J$.

Consequently, if $k \ne 0$, the solution of the problem $(P_{L_0,L_1}[0])$ is a nonpositive function in J and the result is proved.   □

The description of the sets N_T and P_T will be done by the direct study of the expression of Green's function.

To this end, let $M = m^2 > 0$, with $m > 0$ and define the following function:

$$k(m) = (p_0 p_1 - q_0 q_1 m^2) \sin \pi\,(b - a) + m(p_0 q_1 + q_0 p_1) \cos \pi\,(b - a).$$

$$(1.10.2)$$

The characterization of N_T is presented in the following result.

Lemma 1.10.3 ([14, Theorem 3.1]). *Let $\bar{m}$ be the first positive root of the function k given by (1.10.2). Then operator $T_2[M]$ is inverse negative on Z if and only if one of the following conditions hold:*

1. $M \in \left(-\infty, \bar{m}^2\right)$ *and one of this situations holds:*

   (a) $p_0,\ p_1,\ q_0,\ q_1 > 0.$
   (b) $p_0 = 0,\ (p_1 > 0\ and\ q_1 > 0)\ or\ p_1 = 0,\ (p_0 > 0\ and\ q_0 > 0).$
   (c) $q_0 = 0,\ (p_1 > 0\ and\ q_1 > 0)\ or\ q_1 = 0,\ (p_0 > 0\ and\ q_0 > 0).$

2. $p_0 = p_1 = 0$ *and* $M \in (-\infty, 0)$.

3. $q_0 = q_1 = 0$ *and* $M \in \left(-\infty, \left(\frac{\pi}{b-a}\right)^2\right)$.

4. $p_0 = q_1 = 0$ *or* $p_1 = q_0 = 0$ *and* $M \in \left(-\infty, \left(\frac{\pi}{2(b-a)}\right)^2\right)$.

Proof. Assume that $J = [0, 2\pi]$. In this case, Green's function related to this problem is given by

$$g_m(t, s) = -\frac{1}{k(m)} \begin{cases} g(m,s)\, f(m,t) \,, 0 \le s \le t \le 2\pi \\ g(m,t)\, f(m,s) \,, 0 \le t \le s \le 2\pi, \end{cases}$$

with

$$g(m, s) = p_0 \sin ms + m q_0 \cos ms,$$

and

$$f(m, t) = \frac{p_1}{m} \sin m(2\pi - t) + q_1 \cos m(2\pi - t).$$

Moreover, the nonhomogeneous function introduced in (1.10.1) is given by

$$h_m(t, L_0, L_1) = \frac{m\, L_0\, f(m,t) + L_1\, g(m,t)}{k(m)}. \tag{1.10.3}$$

Following the proof of [14, Theorem 3.1] one can check, by studying directly the expression of Green's function g_m, that $g_m \le 0$ on $J \times J$ if and only if $m \in (0, \bar{m})$.

We note that this property can be deduced by applying Lemma 1.8.25. To this end, we must take into account that the supremum of the intervals coincide, in the four situations, with the first eigenvalue of the operator $T_2[M]$ on the set

$$X = \left\{ u \in W^{2,1}(J, \mathbb{R}); \qquad B_0\, u(0) = 0 \quad B_1\, u(2\pi) = 0 \right\}.$$

The second case follows from Lemma 1.10.1 and the fact that if $p_0 = q_0 = 0$, $M = 0$ is an eigenvalue of operator $T_2[M]$.

To prove the rest of the situations we take into account that there is ϕ, a solution of the problem $u'' + \bar{m}^2 u = 0$, $u \in X$, that is strictly positive in $(0, 2\pi)$. It is not difficult to verify that $g(t, s)/\phi(t)$ is bounded in $J \times J$, where g is Green's function related to operator $L[0]$ shown in Lemma 1.10.2. In particular, the condition (N_g) is fulfilled and the negativeness of g_m for $m \in (0, \bar{m})$ holds as a direct consequence of Lemma 1.8.33.

On the other hand, using (1.10.3) it is obvious that if $L_0, L_1 \le 0$, then $h_m(t, L_0, L_1) \le 0$ in J for all $m \in (0, \bar{m})$ and the proof is concluded from the change of variables showed in Lemma 1.4.15. $\qquad \square$

By an exhaustive study of Green's function related to operator $T_2[M]$, the set P_T is characterized in [14] for $J = [0, 2\pi]$. Such result together with the change of variables given in Lemma 1.4.15 read as follows.

Lemma 1.10.4 ([14, Theorem 3.2]). *Let $\bar{m}$ be the first positive root of the function k given by (1.10.2). Then the operator $T_2[M]$ is inverse positive on Z if and only if one of the following conditions hold:*

1. $p_0, p_1, q_0, q_1 > 0$ *and* $M \in (\bar{m}^2, \hat{m}^2]$, *where $\hat{m}$ is the least positive solution of the expression*

$$\tan m (b - a) = -\max\left\{\frac{q_0}{p_0}, \frac{q_1}{p_1}\right\} m.$$

2. $p_0 = p_1 = 0$ *and* $M \in \left(0, \left(\frac{\pi}{2(b-a)}\right)^2\right]$.
3. $p_0 = 0,$ *($p_1 > 0$ and $q_1 > 0$) or $p_1 = 0,$ ($p_0 > 0$ and $q_0 > 0$) and*

$$M \in \left(\bar{m}^2, \left(\frac{\pi}{2(b-a)}\right)^2\right].$$

As a conclusion, the following consequence is attained.

Corollary 1.10.5 ([14, Theorem 3.3]). *If one of the following three conditions holds, then there exists no $M \in \mathbb{R}$ such that the operator $T_2[M]$ is inverse positive on Z.*

1. $q_0 = q_1 = 0.$
2. $p_0 = q_1 = 0$ *or* $p_1 = q_0 = 0.$
3. $q_0 = 0,$ *($p_1 > 0$ and $q_1 > 0$) or $q_1 = 0,$ ($p_0 > 0$ and $q_0 > 0$).*

1.10.2 Neumann Boundary Conditions

This part is devoted to the study of the sets N_T and P_T related to the general second-order operator

$$T_{\gamma,M}\, u(t) = u''(t) + 2\gamma\, u'(t) + M\, u(t)$$

on the set

$$Z_2 = \left\{u \in W^{2,1}(J, \mathbb{R}); \quad u'(a) \geq 0 \geq u'(b)\right\}.$$

Arguing as in the proof of Lemma 1.10.1, it is immediate to verify that if $M < 0$, then $u \geq 0$ on J for every $\gamma \in \mathbb{R}$, that is, $N_T = (-\infty, 0)$.

From Lemma 1.4.15 we can consider $J = [0, R]$, for $R > 0$ given. The results that are presented here are given in [16, Sect. 2] in the set of functions in $\mathscr{C}^2(J, \mathbb{R})$. Since the extension to $W^{2,1}(J, \mathbb{R})$ is immediate, we will work in such space. Thus, let $\sigma \in \mathscr{L}^1(J, \mathbb{R})$ a nonnegative function a.e. in J; we are interested in describing the set of the parameters $\gamma \in \mathbb{R}$ and $M > 0$ for which the following problem has a unique nonnegative solution for every $A \geq 0 \geq B$:

$$u''(t)+2\gamma u'(t)+M u(t) = \sigma(t), \quad \text{a.e. } t \in I, \ u'(0) = A, \quad u'(R) = B. \quad (1.10.4)$$

In this case, instead of obtaining the exact expression of Green's function, we will make a different approach. To this end define the solutions $v_0(t)$ and $v_R(t)$ of the Cauchy problems

$$v_0''(t) + 2\gamma v_0'(t) + M v_0(t) = 0, \ v_0(0) = 1, \quad v_0'(0) = 0 \qquad (1.10.5)$$

and

$$v_R''(t) + 2\gamma v_R'(t) + M v_R(t) = 0, \ v_R(R) = 1, \quad v_R'(R) = 0. \qquad (1.10.6)$$

Now, writing the equation in self-adjoint form

$$(e^{2\gamma t}u'(t))' + M e^{2\gamma t}u(t) = e^{2\gamma t}\sigma(t), \quad t \in I, \qquad (1.10.7)$$

and proceeding in the same way with (1.10.5) and (1.10.6). Multiplying (1.10.7) by v_0 and (1.10.5) by u, integrating by parts and subtracting, we conclude that

$$- e^{2\gamma R}v_0'(R) u(R) = \int_0^R e^{2\gamma t}\sigma(t) v_0(t) \, dt - e^{2\gamma R}v_0(R) B + A. \qquad (1.10.8)$$

In the same way, using (1.10.6) we obtain

$$v_R'(0) u(0) = \int_0^R e^{2\gamma t}\sigma(t) v_R(t) \, dt - e^{2\gamma R}B + v_R(0) A. \qquad (1.10.9)$$

It is obvious that the condition

$$v_0(t) > 0 \text{ in } [0, R), \ v_0'(R) < 0, \ v_R(t) > 0 \text{ in } (0, R], \ v_R'(0) > 0, \qquad (1.10.10)$$

implies that $u(0) \geq 0$ and $u(R) \geq 0$ whenever $\sigma(t) \geq 0$ a.e. in J and $B \leq 0 \leq A$. Moreover, the inequalities are strict if $\sigma \succ 0$ in J. In fact, (1.10.10) also implies that (1.10.4) is uniquely solvable.

In the next result we study the zeroes of functions v_0 and v_R.

Lemma 1.10.6 ([16, Proposition 2.1]). *The following assertions hold:*

(a) Let $0 < M < \gamma^2$. Then v_0 has exactly one zero $y(M, \gamma)$ given by the expression

$$y(M, \gamma) = -\text{sign}(\gamma) \frac{\log (|\gamma| + \sqrt{\gamma^2 - M}) - \log (|\gamma| - \sqrt{\gamma^2 - M})}{2\sqrt{\gamma^2 - M}},$$
$$(1.10.11)$$

which satisfies $v_0'(y(M, \gamma)) \text{sign}(\gamma) > 0$.

(b) *Let* $0 < \gamma^2 = M$. *Then* v_0 *has exactly one zero* $y(M, \gamma) = -1/\gamma$. *Furthermore, it is satisfied that* $v_0'(y(M, \gamma)) \operatorname{sign}(\gamma) > 0$.

(c) *Let* $0 < \gamma^2 < M$. *Then* v_0 *has a smallest positive zero* $y_1(M, \gamma)$ *and a largest negative zero* $y_2(M, \gamma)$ *given by*

$$y_1(M, \gamma) = \frac{1}{\sqrt{M - \gamma^2}} \left(\frac{\pi}{2} + \operatorname{sign}(\gamma) \arctan \left(\frac{|\gamma|}{\sqrt{M - \gamma^2}} \right) \right), \qquad (1.10.12)$$

$$y_2(M, \gamma) = -\frac{1}{\sqrt{M - \gamma^2}} \left(\frac{\pi}{2} - \operatorname{sign}(\gamma) \arctan \left(\frac{|\gamma|}{\sqrt{M - \gamma^2}} \right) \right). \qquad (1.10.13)$$

In this case, $v_0'(y_2(M, \gamma)) > 0$ *and* $v_0'(y_1(M, \gamma)) < 0$ *hold.*

Proof. (a) If $\gamma > 0$, $v_0(t) = (s\, e^{rt} - r\, e^{st})/(s - r)$ where $r = -\gamma - \sqrt{\gamma^2 - M}$ and $s = -\gamma + \sqrt{\gamma^2 - M}$. We obtain the unique zero of v_0 as $t = (\log|r| - \log|s|)/(r - s)$. The remaining assertion follows immediately from the formula for v_0.

If $\gamma < 0$, just note that (with obvious notation) $v_0(t; \gamma) = v_0(-t; -\gamma)$.

(b) If $M = \gamma^2$, then $v_0(t) = e^{-\gamma t}(1 + \gamma\, t)$ and the conclusion holds trivially.

(c) If $\gamma > 0$, then for some $C > 0$

$$v_0(t) = C\, e^{-\gamma t} \cos \sqrt{M - \gamma^2}\, (t - \Phi),$$

where $\Phi = \dfrac{1}{\sqrt{M-\gamma^2}} \arctan \dfrac{\gamma}{\sqrt{M-\gamma^2}}$. The zeros of v_0 are then given by

$$\frac{1}{\sqrt{M - \gamma^2}} \left(\arctan \frac{\gamma}{\sqrt{M - \gamma^2}} + (2n + 1)\frac{\pi}{2} \right), \quad n \in \mathbb{Z}.$$

Taking $n = 0, -1$ we obtain y_1 and y_2.

If $\gamma < 0$, conclude as in (a). $\square$

Remark 1.10.7. (i) Since $v_R(t) = v_0(t - R)$ we conclude that the expressions in the right-hand sides of (1.10.11)–(1.10.13) also give, in each case, the distance to R of zeros of v_R that are nearest R.

(ii) It is clear that in case (a) $v_0'(t) \neq 0$ if $t \neq 0$ and in case (c) $v_0'(t) \neq 0$ if $t \in (y_2, 0) \cup (0, y_1)$. A similar remark applies to v_R and justifies the statement before Proposition 1.10.6.

From Lemma 1.10.6 and Remark 1.10.7, we arrive at the following corollary.

Corollary 1.10.8 ([16, Corollary 2.1]). *The following assertions hold:*

(a) If $0 < M < \gamma^2$, *condition (1.10.10) is satisfied if and only if*

$$\phi(M,\gamma) \equiv \frac{\log{(|\gamma| + \sqrt{\gamma^2 - M})} - \log{(|\gamma| - \sqrt{\gamma^2 - M})}}{2\sqrt{\gamma^2 - M}} \geq R.$$

$$(1.10.14)$$

(b) If $0 < \gamma^2 = M$, *condition (1.10.10) is satisfied if and only if* $R \leq 1/|\gamma|$.
(c) If $0 < \gamma^2 < M$, *condition (1.10.10) is satisfied if and only if*

$$\psi(M,\gamma) \equiv \frac{1}{\sqrt{M - \gamma^2}}\left(\frac{\pi}{2} - \arctan\frac{|\gamma|}{\sqrt{M - \gamma^2}}\right) \geq R. \qquad (1.10.15)$$

By means of a careful analysis of the functions ϕ and ψ we deduce the following results.

Lemma 1.10.9 ([16, Proposition 2.2]). *The functions ϕ and ψ defined in the previous corollary verify the following properties:*

(a) The function $\phi(\cdot,\gamma)$ defined in (1.10.14) is strictly decreasing in $(0,\gamma^2)$ with $\phi(0^+,\gamma) = +\infty$, $\phi(\gamma^{2-},\gamma) = 1/|\gamma|$.
(b) The function $\psi(\cdot,\gamma)$ defined in (1.10.15) is strictly decreasing in $(\gamma^2,+\infty)$ with $\psi(\gamma^{2+},\gamma) = 1/|\gamma|$, $\psi(+\infty,\gamma) = 0$.

Lemma 1.10.10 ([16, Proposition 2.3]). *Define*

$$\theta(M,\gamma) = \begin{cases} \phi(M,\gamma) & \text{if } 0 < M < \gamma^2 \\ 1/|\gamma| & \text{if } M = \gamma^2 \\ \psi(M,\gamma) & \text{if } 0 < \gamma^2 < M. \end{cases}$$

Then $\theta(\cdot,\gamma)$ is continuous and strictly decreasing in $(0,+\infty)$ with $\theta(0^+,M) = +\infty$, $\theta(+\infty,\gamma) = 0$ and condition (1.10.10) is equivalent to $\theta(M,\gamma) \geq R$.

Remark 1.10.11 ([16, Remark 2.2]).

(a) The equation $\theta(M,\gamma) = R$ defines a function $M \equiv M(R,\gamma)$ strictly decreasing with respect to $R > 0$. In fact this is a $\mathscr{C}^1$-function (a simple calculation shows that $\frac{\partial\theta}{\partial M}(\gamma^2,\gamma) = -\frac{1}{3|\gamma|^3}$). The construction of θ also shows that if the first eigenvalue of the problem

$$v''(t) + 2\gamma v'(t) + M v(t) = 0, \quad t \in I, \quad v(0) = 0, \quad v'(R) = 0, \quad (1.10.16)$$

(respectively: $v''(t) + 2\gamma v'(t) + M v(t) = 0, v'(0) = 0, v(R) = 0$) is denoted by $M_+(R,\gamma)$ (resp. $M_-(R,\gamma)$), then we have $M(R,\gamma) = M_+(R,\gamma) < M_-(R,\gamma)$ if $\gamma > 0$, $M(R,\gamma) = M_-(R,\gamma) < M_+(R,\gamma)$ if $\gamma < 0$.

(b) If $\theta(M, \gamma) < R$ and M is near $M(R, \gamma)$ formulas (1.10.8) and (1.10.9) show
    that there exist positive functions σ such that $u(0) < 0$ or $u(R) < 0$.

So we arrive at the main result of this part, in which the set P_T is described.

Theorem 1.10.12 ([16, Proposition 2.4]). *Operator $T_{\gamma,M}$ is inverse positive on Z_2 if and only if*

$$0 < M \le \frac{M(1, \gamma\,(b - a))}{(b - a)^2},$$

with function M defined on Remark 1.10.11.

Proof. Suppose, in a first moment, that $J = [0, R]$. Set $M_1 = M(R, \gamma)$. Let $\sigma \in \mathcal{L}^1(J)$, $\sigma \succ 0$ on J. Let u be the unique solution of problem (1.10.4), supposing that it is not positive on J. Assume also (according to remark (a) above) that $M_1 = M_+(R, \gamma)$. By our assumption there exists $t_1 \in (0, R)$ such that $u(t_1) = 0$ and $u > 0$ in $(t_1, R]$. Since (1.10.16) has a positive solution $v(t)$ in $(0, R]$ with $M = M_1$, we obtain

$$\left[e^{2\gamma t}(u'(t)v(t) - u(t)v'(t)) \right]_{t_1}^{R} + (M - M_1)\int_{t_1}^{R} e^{2\gamma t}\,u(t)\,v(t)\,dt = \int_{t_1}^{R} e^{2\gamma t}\,\sigma(t)\,v(t)\,dt.$$

Since $u'(t_1) \ge 0$ and $v(t_1) > 0$, this identity implies $M > M_1$ (note that $u'(t_1) = 0$ and $\sigma(t) \equiv 0$ for all $t \in [t_1, R]$ cannot hold simultaneously; since then we would have $u(t) \equiv 0$ in $[t_1, R]$). A similar argument applies in case $M_1 = M_-(R, \gamma)$.

As a consequence we have proved that $u > 0$ on J for all $0 < M \le M(R, \gamma)$. As a consequence the related Green's function is nonnegative on $J \times J$.

Notice that since for any $s \in (0, T)$ given, Green's function $z(\cdot) = g(\cdot, s)$ satisfies that $T_{\gamma,M}\, z(t) = 0$ for all $t \in [0, s) \cup (s, R]$, we have that there is no $t_0 \in J$ such that $z(t_0) = z'(t_0)$. In particular, if $g \ge 0$ on $J \times J$, then it can vanish only at the diagonal.

Remark 1.10.11 together with Theorem 1.8.9 imply that such estimate is optimal on $[0, R]$.

Now, Lemma 1.4.15 tells us that the sign of Green's function related to operator $T_{\gamma,M}$ on the space

$$X_2 = \left\{ u \in W^{2,1}(J, \mathbb{R}); \quad u'(a) = u'(b) = 0 \right\}$$

coincides with the one of operator $T_{\gamma\,(b-a), M\,(b-a)^2}$ on the space

$$X_2 = \left\{ u \in W^{2,1}(J, \mathbb{R}); \quad u'(0) = u'(1) = 0 \right\}.$$

But in this second situation, Green's function is nonnegative on $[0, 1] \times [0, 1]$ if and only if

$$0 < M\,(b - a)^2 \le M(1, \gamma\,(b - a))$$

and the proof is concluded. $\square$

Remark 1.10.13. Note that from the arguments given in the previous proof, by using Lemma 1.4.15, it is not difficult to verify that the function M introduced in Remark 1.10.11 satisfies that

$$M(R, \gamma) = \frac{M(1, \gamma R)}{R^2}.$$

Remark 1.10.14. It is obvious that the function ψ defined in (1.10.15) satisfies that $\psi(M, 0) = \frac{\pi}{2\sqrt{M}}$.

In particular $\psi(M, 0) = R$ if and only if $M = \left(\frac{\pi}{2R}\right)^2 \equiv M(R, 0) = \frac{M(1,0)}{R^2}$.

As consequence, we deduce, from Theorem 1.10.12, that operator $u'' + M u$ is inverse positive on Z_2 if and only if

$$0 < M \le \left(\frac{\pi}{2(b-a)}\right)^2$$

as we know from Lemma 1.10.4.

1.10.3   Simply Supported Beam Conditions

This subsection is devoted to the study of the fourth-order linear operator $T_4[M] u = u^{(4)} + M u$ coupled with the so-called simply supported beam conditions defined in the space

$$X_4 = \{u \in W^{4,1}(J), \quad u(a) = u(b) = u''(a) = u''(b) = 0\}.$$

It is not difficult to verify that the homogeneous problem

$$T_4[M] u(t) = 0, \ t \in J, \quad u \in X_4,$$

has nontrivial solutions if and only if

$$M = -\left(\frac{k\pi}{b-a}\right)^4, \quad k = 1, 2, \ldots$$

Moreover, it is a self-adjoint operator and, as a consequence, the structure of N_T and P_T has no direct relationship except for the one given in Theorem 1.8.35.

In Example 1.8.28 we have shown that operator $T_4[0]$ satisfies condition (P_g). Thus, Theorem 1.8.31 ensures that

$$\left(-\left(\frac{\pi}{b-a}\right)^4, 0\right] \subset P_T.$$

The set P_T is completely described in [50], Chap. 2, Sect. 4.1.3. The result is the following.

Lemma 1.10.15. *Operator $T_4[M]$ is inverse positive in X_4 if and only if*

$$M \in \left(-\left(\frac{\pi}{b-a}\right)^4, 4\left(\frac{k_0}{b-a}\right)^4\right].$$

Here $k_0 \approx 3.9266$ is the smallest positive solution of the equation $\tan k = \tanh k$.

Proof. The case $M < 0$ has been proved above. Let $M = m^4 > 0$ for some $m > 0$ and $J = [0, 1]$. Denote $\bar{m} = \sqrt{2}\,k_0$.

Green's function $g_m(t, s)$ can be calculated explicitly (see Appendix B). One can see that $g_{\bar{m}}(t, s) \geq 0$ on $J \times J$. So Theorem 1.8.9 implies that the same holds for all $m \in [0, \bar{m}]$.

To see that such estimation is optimal, it is not difficult to verify that $\Phi_m(t) = g_m(t, 0)$ satisfies

$$T_4[m^4]\,\Phi_m(t) = 0, \quad t \in J, \quad \Phi_m(0) = 0, \quad \Phi'_m(0) = -1, \quad \Phi_m(1) = 0, \quad \Phi''_m(1) = 0.$$

By elementary calculations one obtains

$$\operatorname{sign}\Phi'(1) = -\operatorname{sign}(\bar{m} - m)$$

for $0 \leq m \leq \bar{m} + \epsilon$ with sufficiently small $\epsilon > 0$. Consequently, Φ assume negative values for m in a right-hand neighborhood of $\bar{m}$. As a consequence, Theorem 1.8.9 implies that g_m changes sign for all $m > \bar{m}$.   □

Remark 1.10.16. In [50] the proof has been done for the interval $[0, 1]$ and functions in $\mathscr{C}^4([0, 1], \mathbb{R})$. Since the regularity of the solutions has no influence on the expression of Green's function the application to X_4, by means of Lemma 1.4.15, is immediate.

The set N_T is described in the following result.

Lemma 1.10.17 ([18, Propositions 2.1 and 3.1]). *Operator $T_4[M]$ is inverse negative in X_4 if and only if*

$$M \in \left[-\left(\frac{k_0}{b-a}\right)^4, -\left(\frac{\pi}{b-a}\right)^4\right).$$

with k_0 defined as in Lemma 1.10.15.

Proof. Consider $J = [0, 1]$ and $M = -m^4$. In such a case Green's function is given by

$$g_m(t,s)=\begin{cases} \dfrac{\csc(m)\sin(m-mt)\sin(ms)-\operatorname{csch}(m)\sinh(m-mt)\sinh(ms)}{2m^3}, & 0 \leq s \leq t \leq 1, \\[3mm] \dfrac{\csc(m)\sin(m-ms)\sin(mt)-\operatorname{csch}(m)\sinh(m-ms)\sinh(mt)}{2m^3}, & 0 \leq t \leq s \leq 1. \end{cases}$$

$$(1.10.17)$$

We shall prove that if $\pi < m \leq k_0$, then $g_m(t, s) < 0$ for all t, $s \in (0, 1)$. From the fact that $k_0 < 2\pi$, we have that $\csc(m) < 0$, so since Green's function g_m is symmetric and $\sinh(m) > 0$, we only must show that for all t, $s \in (0, 1)$

$$\sin(mt)\sin(m(1-s))\sinh(m) - \sin(m)\sinh(m(1-s))\sinh(mt) > 0,$$

which making $\tau = 1 - s$ is equivalent to

$$\frac{\sin(mt)\sin(m\tau)}{\sin(m)} < \frac{\sinh(mt)\sinh(m\tau)}{\sinh(m)} \qquad \text{for all } t,\ \tau \in (0, 1). \qquad (1.10.18)$$

Clearly it suffices to consider the case $\sin(m\tau) > 0$ and $\sin(mt) < 0$. Since $\sin(x) < \sinh(x)$ for all $x > 0$ it is enough to prove that

$$\frac{\sin(mt)}{\sinh(mt)} > \frac{\sin(m)}{\sinh(m)} \qquad \text{for all } t \in (0, 1). \qquad (1.10.19)$$

But this inequality follows immediately from the fact that the derivative of $\dfrac{\sin(x)}{\sinh(x)}$ is strictly negative in $(0, k_0)$. Therefore, since $mt < m \leq k_0$, we have that (1.10.19) holds.

To verify that this estimation is optimal it suffices to take into account that

$$\frac{d}{dt}g_m(t, t) = \frac{\csc(m)\sin(m(1-2t)) - \operatorname{csch}(m)\sinh(m(1-2t))}{2m^2}.$$

It is easy to verify, using the fact that $\frac{\sin x}{\sinh x}$ becomes increasing in a right neighborhood of k_0, that there exist $\epsilon > 0$ such that $\frac{d}{dt}g_m(t, t) > 0$ for all $t \in (0, \epsilon)$ and $m \in (k_0, k_0 + \epsilon)$.

From Theorem 1.8.9 and Lemma 1.4.15 we conclude the proof. $\qquad\qquad\square$

In the sequel some anti-maximum principles are shown for operator $T_4[M]$ in bigger sets than X_4. To this end, in the next result, we present the general solution of the nonhomogeneous boundary value problem.

Lemma 1.10.18. *Let $\sigma \in \mathscr{L}^1(J, \mathbb{R})$ and α, β, γ, $\delta \in \mathbb{R}$ be fixed. Assume that operator $T_4[M]$ is invertible in X_4, then the unique solution of problem*

$$T_4[M]u(t) = \sigma(t),\ \text{a.e.}\ \ t \in I,\ \ u(a) = \alpha,\ u(b) = \beta,\ u''(a) = \gamma,\ u''(b) = \delta,$$

is given by the following expression:

$$u(t) = (b-a)^3 \int_a^b g_m\left(\frac{t-a}{b-a}, \frac{s-a}{b-a}\right)\sigma(s)\,ds$$
$$+\alpha\, x_m(t) + \beta\, x_m(a+b-t) + \gamma\, y_m(t) + \delta\, y_m(a+b-t), \qquad (1.10.20)$$

where we denote $M = \pm m^4$ (depending on the sign of M), g_m is given in (1.10.17) if $M < 0$, and in Appendix B if $M > 0$, x_m, and y_m are defined respectively as the unique solutions of the following problems:

$$T_4[M]\, w(t) = 0, \ a.e. \quad t \in J, \quad w(a) = 1, \ w(b) = w''(a) = w''(b) = 0,$$

and

$$T_4[M]\, w(t) = 0, \ a.e. \quad t \in J, \quad w(a) = w(b) = 0, \ w''(a) = 1, \ w''(b) = 0.$$

Next, we shall prove different maximum principles for the case $M \geq 0$.

Lemma 1.10.19 ([18, Theorem 2.1]). *Let $M \geq 0$. Then the linear operator $T_4[M]$ is inverse positive in the space*

$$W_1 = \{u \in W^{4,1}(J) : u(a) \geq 0, \ u(b) \geq 0, \ u''(a) = u''(b) = 0\}$$

if and only if

$$M \in \left[0, \ 4\left(\frac{\pi\, M_4}{b - a}\right)^4\right],$$

where M_4 is defined as in Lemma 1.9.17.

Proof. Assume that $J = [0, 1]$. Denote $k_1 := M_4\,\pi$. By definition of M_4, k_1 is the smallest positive solution of the equation $\tan k = -\tanh k$. Since $k_1 < k_0$, from expression (1.10.20) we have that this result is true if and only if $x_m \geq 0$.

One can verify, by explicit calculation, that function x_m, is given for $m > 0$ by

$$x_m(t) = \frac{-\cos\left(\frac{mt}{\sqrt{2}}\right)\cosh\left(\frac{m(t-2)}{\sqrt{2}}\right) + \cos\left(\frac{m(t-2)}{\sqrt{2}}\right)\cosh\left(\frac{mt}{\sqrt{2}}\right)}{\cos\left(\sqrt{2}m\right) - \cosh\left(\sqrt{2}m\right)} \qquad (1.10.21)$$

and $x_0(t) = 1 - t$.

Let us see that $x_m \geq 0$ in $[0, 1]$ if and only if $m \in [0, \sqrt{2}k_1]$.

First we observe that x_m cannot have a double zero in $(0, 1)$. Indeed, since x_m is the minimizer of the functional

$$\int_0^1 \left(w''^2(s) + M w^2(s)\right) ds$$

in $\mathscr{W}^{2,2}(J, \mathbb{R})$ coupled with the boundary conditions $w(0) = 1$ and $w(1) = 0$. So, if $t_0 \in (0, 1)$ is a double zero of x_m, then $x_m(t) = 0$ for all $t \in [t_0, 1]$, which is impossible.

Next we remark that

$$x'_m(1) = \frac{\sqrt{2}m\left(\cosh\left(\frac{m}{\sqrt{2}}\right)\sin\left(\frac{m}{\sqrt{2}}\right) + \cos\left(\frac{m}{\sqrt{2}}\right)\sinh\left(\frac{m}{\sqrt{2}}\right)\right)}{\cos\left(\sqrt{2}m\right) - \cosh\left(\sqrt{2}m\right)},$$

from which we conclude that $x'_m(1) < 0$ for all $0 < m < \sqrt{2}k_1$, being $\sqrt{2}k_1$ the first positive zero of the equation $x'_m(1) = 0$.

Now suppose that for some $0 < m \leq \sqrt{2}k_1$ the function x_m takes negative values. Using a continuity argument and taking the infimum of such values of $m > 0$, we obtain a $\bar{m} \in (0, \sqrt{2}k_1)$ such that $x_{\bar{m}}$ has a double zero which is different from 1 since $x'_{\bar{m}}(1) < 0$, but this is a contradiction.

Let $m > \sqrt{2}k_1$ be fixed. We shall prove that x_m has a zero in $(0, 1)$ and, since the zero must be simple, x_m changes sign. In view of (1.10.21) we have that $x_m(t) = 0$ if and only if $h(t) := f(t) - f(t-2) = 0$ where

$$f(t) = \frac{\cos\left(\frac{mt}{\sqrt{2}}\right)}{\cosh\left(\frac{mt}{\sqrt{2}}\right)}.$$

It is easy to see that $\frac{\cos s}{\cosh s}$ has its unique absolute maximum at $s = 0$ and an absolute minimum at $s = k_1$. Therefore $h(0) > 0$ and $h(\frac{\sqrt{2}k_1}{m}) \leq 0$. Since $\frac{\sqrt{2}k_1}{m} < 1$ the result follows from Bolzano's theorem. The estimate for any interval $[a, b]$ holds from Lemma 1.4.15. □

When function u attains the same nonnegative value at the endpoints of the interval the set of M for which the operator is inverse positive can be enlarged.

Lemma 1.10.20 ([18, Theorem 2.2]). *Let $M \geq 0$. Then the linear operator $T_4[M]$ is inverse positive in the space*

$$W_2 = \{u \in W^{4,1}(J) : u(a) = u(b) \geq 0, \ u''(a) = u''(b) = 0\}$$

if and only if

$$M \in \left[0, 4\left(\frac{\pi}{b-a}\right)^4\right].$$

Proof. Consider again $J = [0, 1]$. Since $0 \leq M \equiv m^4 \leq 4\pi^4 < 4k_0^4$, it follows from Lemma 1.10.15 that $g_m > 0$ in $(0, 1) \times (0, 1)$. Thus, by using equation (1.10.20), it is enough to prove that function

$$w_m(t) = x_m(t) + x_m(1 - t)$$

is nonnegative in J if and only if $m \in [0, \sqrt{2}\pi]$.

One can verify, by explicit calculation, that $w_0 \equiv 1$ and, for all $m > 0$, function w_m is given by the following expression:

$$w_m(t) = \left[-\cos\left(\frac{mt}{\sqrt{2}}\right) \cosh\left(\frac{m(t-2)}{\sqrt{2}}\right) + \cos\left(\frac{m(t+1)}{\sqrt{2}}\right) \cosh\left(\frac{m(t-1)}{\sqrt{2}}\right) \right.$$
$$\left. + \cos\left(\frac{m(t-2)}{\sqrt{2}}\right) \cosh\left(\frac{mt}{\sqrt{2}}\right) - \cos\left(\frac{m(t-1)}{\sqrt{2}}\right) \cosh\left(\frac{m(t+1)}{\sqrt{2}}\right) \right]$$
$$/ \left[\cos\left(\sqrt{2}m\right) - \cosh\left(\sqrt{2}m\right) \right]$$

Claim 1. If $w_m(t) \geq 0$ for all $t \in J$, then $\min_{t \in J} w_m(t) = w_m(1/2)$.

From the definition, it is obvious that function w_m is symmetric with respect to $t = 1/2$. Moreover, if $w_m(t) \geq 0$, then w_m'' is concave and then using the boundary conditions we conclude that w_m is convex. Therefore the claim follows.

Claim 2. The set $A = \{m \geq 0 : w_m \geq 0 \ \text{ in } J\}$ is an interval.

Clearly A is nonempty because $0 \in A$. If $m_1 \in [0, \sqrt{2}\,k_0]$ is such that $m_1 \in A$ and $0 \leq m_2 < m_1$, then $m_2 \in A$. Indeed, from the equations

$$T_4[m_i^4]\, w_i = 0, \ t \in J, \quad w_i(0) = w_i(1) = 1, \ w_i''(0) = w_i''(1) = 0,$$

for $i = 1, 2$, it follows that $w = w_2 - w_1$ satisfies

$$T_4[m_2^4]\, w = (m_1^4 - m_2^4)w_1 \succ 0 \ \text{ on } J, \quad w(0) = w(1) = 0, \ w''(0) = w''(1) = 0.$$

Now, Lemma 1.10.15 ensures that $w_2 > w_1$ in $(0, 1)$ and, as a consequence, $m_2 \in A$.

Since

$$w_m\left(\frac{1}{2}\right) = \frac{2\cos\left(\frac{m}{2\sqrt{2}}\right) \cosh\left(\frac{m}{2\sqrt{2}}\right)}{\cos\left(\frac{m}{\sqrt{2}}\right) + \cosh\left(\frac{m}{\sqrt{2}}\right)},$$

the preceding argument shows that the conditions $m_1 > \sqrt{2}k_0$ and $m_1 \in A$ are impossible; otherwise we obtain that $m_2 \in A$ for all $m_2 \in (\sqrt{2}\pi, \sqrt{2}k_0)$ but $w_{m_2}(1/2) < 0$ for all $m_2 \in (\sqrt{2}\pi, \sqrt{2}k_0)$. Therefore A is an interval contained in $[0, \sqrt{2}\,k_0]$.

Claim 3. $A = [0, \sqrt{2}\,\pi]$

By continuity A is a closed interval $[0, l]$. If $l < \sqrt{2}\,\pi$, then $w_l(t) \geq w_l(1/2) > 0$ for all $t \in I$, and again by continuity $w_m \geq 0$ for all m in a small enough right neighborhood of l, a contradiction. On the other hand, since $w_m(1/2) < 0$ in a right neighborhood of $\sqrt{2}\,\pi$, we have that $l \leq \sqrt{2}\,\pi$. Lemma 1.4.15 gives us the best value in $[a, b]$.   □

Lemma 1.10.21 ([18, Theorem 2.3]). *Let $M \geq 0$. Then the linear operator $T_4[M]$ is inverse positive in the space*

$$W_3 = \{u \in W^{4,1}(J) : u(a) = u(b) = 0, \ u''(a) \leq 0, \ u''(b) \leq 0\}$$

if and only if

$$M \in \left[0, 4 \left(\frac{k_0}{b-a}\right)^4\right],$$

with k_0 defined as in Lemma 1.10.15.

Proof. In this case, the result holds if and only if $g_m \geq 0$ on $J \times J$ and $y_m \geq 0$ on J.

It is enough to take into account that $(J = [0, 1])$

$$y_m(t) = \frac{-\sin\left(\frac{mt}{\sqrt{2}}\right) \sinh\left(\frac{m(t-2)}{\sqrt{2}}\right) + \sin\left(\frac{m(t-2)}{\sqrt{2}}\right) \sinh\left(\frac{mt}{\sqrt{2}}\right)}{m^2 \left(\cos\left(\sqrt{2}m\right) - \cosh\left(\sqrt{2}m\right)\right)},$$

$$y'_m(1) = \frac{\sqrt{2}e^{-\frac{m}{\sqrt{2}}}\left(\left(1 - e^{\sqrt{2}m}\right)\cos\left(\frac{m}{\sqrt{2}}\right) + (1 + e^{\sqrt{2}m})\sin\left(\frac{m}{\sqrt{2}}\right)\right)}{m\left(1 + e^{2\sqrt{2}m} - 2e^{\sqrt{2}m}\cos\left(\sqrt{2}m\right)\right)}$$

and use similar arguments to those in Theorem 1.10.19.   □

As it is pointed out in [18, Remark 2.3] in this situation it is not considered the case $u''(a) = u''(b) \geq 0$, because the obtained estimate is the same as in Lemma 1.10.15 and the result cannot be improved.

As a direct consequence of Lemmas 1.10.19 and 1.10.21 the following result is obtained.

Corollary 1.10.22 ([18, Corollary 2.1]). *Let $M \geq 0$. Then the linear operator $T_4[M]$ is inverse positive in the space*

$$W_4 = \{u \in W^{4,1}(J) : u(a) \geq 0, \; u(b) \geq 0, \; u''(a) \leq 0, \; u''(b) \leq 0\}$$

if and only if

$$M \in \left[0, 4 \left(\frac{\pi M_4}{b-a}\right)^4\right],$$

where M_4 is defined as in Lemma 1.9.17.

The next consequence follows immediately from Lemmas 1.10.20 and 1.10.21.

Corollary 1.10.23 ([18, Corollary 2.1]). *Let $M \geq 0$. Then the linear operator $T_4[M]$ is inverse positive in the space*

$$W_5 = \{u \in W^{4,1}(J) : u(a) = u(b) \geq 0, \; u''(a) \leq 0, \; u''(b) \leq 0\}$$

if and only if

$$M \in \left[0, 4 \left(\frac{\pi}{b-a}\right)^4\right].$$

For the case $M = -m^4 < 0$ a detailed analysis of the functions x_m, w_m, and y_m discloses that x_m and w_m always change sign and $y_m \geq 0$ in J if and only if $m \in (\pi/(b-a), k_0/(b-a)]$. The following result is shown in [18, Theorem 2.4].

Lemma 1.10.24. *Let*

$$M \in \left[-\left(\frac{k_0}{b-a}\right)^4, -\left(\frac{\pi}{b-a}\right)^4\right).$$

with k_0 defined as in Lemma 1.10.15.

Then the linear operator $T_4[M]$ is inverse negative in the space W_3 defined as in Lemma 1.10.21.

1.10.4   Clamped Beam Conditions

This subsection is devoted to the study of the fourth-order operator $T_4[M]u = u^{(4)} + M u$ in the space of the Clamped beam conditions.

$$Y_4 = \{u \in W^{4,1}(J), \quad u(a) = u(b) = u'(a) = u'(b) = 0\}.$$

We will characterize the set P_T of the real parameters M for which the operator is inverse positive. If $M < 0$, we make use again of the classical Krein-Rutman theorem (Theorem 1.8.16).

Along this part, we will assume $J = [0, 1]$. Adapting the estimations to any arbitrary interval is immediate from Lemma 1.4.15.

Let $M = -m^4 < 0$ for some $m > 0$ be given, and consider the boundary value problem

$$T_4[-m^4]u(t), \; t \in [0, 1]; \quad u(0) = u'(0) = u(1) = u'(1) = 0. \qquad (1.10.22)$$

It is not difficult to verify that this problem has a nontrivial solution if and only if m solves the equation

$$\cos(m)\cosh(m) = 1. \qquad (1.10.23)$$

Moreover, the first positive root of (1.10.23) is $m_1 \approx 4,73004$.

It is not difficult to verify that the function

$$\phi(x) = \frac{1}{C_1}\left(\sin m_1 x - \sinh m_1 x + (\cosh m_1 x - \cos m_1 x)\frac{\sin m_1 - \sinh m_1}{\cos m_1 - \cosh m_1}\right),$$

with

$$C_1 = \frac{-\cos\left(\frac{m_1}{2}\right)\sinh\left(\frac{m_1}{2}\right) + \sin\left(\frac{m_1}{2}\right)\cosh\left(\frac{m_1}{2}\right)}{\cos\left(\frac{m_1}{2}\right) + \cosh\left(\frac{m_1}{2}\right)},$$

is a nontrivial eigenfunction of problem (1.10.22) related to the eigenvalue m_1. This function is positive in $(0, 1)$, $\|\phi\|_\infty = 1$ and, moreover,

$$\phi''(0) = \phi''(1) = \frac{2\,m_1^2\,(\sinh m_1 - \sin m_1)}{C_1(\cos m_1 - \cosh m_1)} > 0.$$

For $M = 0$, by direct integration, one can deduce that Green's function related to operator $T_4[0]$ follows the expression

$$g_0(t, s) = -\frac{1}{6}\begin{cases} s^2\,(t-1)^2\,(s - 3t + 2st), & \text{if } 0 \le s \le t \le 1, \\ t^2\,(s-1)^2\,(t - 3s + 2st), & \text{if } 0 \le t < s \le 1. \end{cases}$$

From the general theory developed in Sect. 1.4, we know that $g_0 \in \mathscr{C}^3([0,1] \times [0,1])$. The boundary conditions coupled with the self-adjoint property of the operator imply that $g_0(t, 0) = g_0(t, 1) = g_0(0, s) = g_0(1, s) = 0$ for all t, $s \in [0, 1]$. Moreover, it is immediate to verify that $g_0(t, s) > 0$ for all $(t, s) \in (0, 1) \times (0, 1)$.
    Furthermore, due to the fact that

$$\lim_{t \to 0^+} \frac{g_0(t, s)}{\phi(t)} = \frac{s\,(s-1)^2}{\phi''(0)} \quad \text{and} \quad \lim_{t \to 1^-} \frac{g_0(t, s)}{\phi(t)} = \frac{s^2\,(1-s)}{\phi''(1)},$$

we conclude that condition (P_g) is fulfilled for $M = 0$.
    As a consequence, Lemma 1.8.33 ensures that operator $T_4[M]$ is inverse positive for all $M \in (-m_1^4, 0]$ and that $-m_1^4 = \inf\{P_T\}$.
    In [12] the same conclusion was obtained without verifying condition (P_g). In that case a Banach space with a weighted norm was defined that provides to the related cones a nonempty interior. The reader can consult this different way that gives us another point of view in order to obtain maximum and anti-maximum principles for linear operators.
    The case $M = m^4 > 0$ for some $m > 0$ has been studied in [12]. In this case the concept of interval of nonoscillation is used, i.e., we say that $[0, 1]$ is an interval of nonoscillation for the differential equation $u^{(4)} + m^4 u = 0$ if no nontrivial solution of the equation $u^{(4)} + m^4 u = 0$ has more than three zeros in $[0, 1]$ (the definition of interval of oscillation can be set as the opposite, that is, there exists u solving $u^{(4)} + m^4 u = 0$ with at least four zeros in the given interval).
    In [49], Schröder proved that if $[0, 1]$ is an interval of nonoscillation for the differential equation $u^{(4)} + m^4 u = 0$, then operator $T_4[M]$ is inverse positive on Y_4. So, the rest of this subsection is devoted to study the values of m for which $[0, 1]$ is an interval of nonoscillation for this operator.

It is clear that the solutions of the fourth-order linear homogeneous equation

$$u^{(4)}(x) + m^4 u(x) = 0, \qquad x \in \mathbb{R}, \tag{1.10.24}$$

are given by the following expression:

$$u(x) = e^{\frac{mx}{\sqrt{2}}}\left(A\cos\left(\frac{mx}{\sqrt{2}}\right) + B\sin\left(\frac{mx}{\sqrt{2}}\right)\right) + e^{-\frac{mx}{\sqrt{2}}}\left(C\cos\left(\frac{mx}{\sqrt{2}}\right) + D\sin\left(\frac{mx}{\sqrt{2}}\right)\right),$$

$$\tag{1.10.25}$$

with $A, B, C, D \in \mathbb{R}$.

Since the equation is autonomous we have that u is a solution of (1.10.24) with $u(x_0) = 0$ if and only if $v(x) = u(x - x_0)$ is also a solution of (1.10.24) and with $v(0) = 0$. In particular, it is enough to look for the solutions of (1.10.24) that vanish at $x = 0$, i.e., we can take $C = -A$ in (1.10.25).

Next , we prove that the nonoscillation property holds in an interval of m.

Lemma 1.10.25 ([12, Lemma 3.1]). *If $[0, 1]$ is an interval of oscillation of (1.10.24) for a given m_*, then it is also an interval of oscillation for all $m > m_*$.*

Proof. Suppose that u is a solution of $u^{(4)} + m_*^4 u = 0$ and $u(0) = u(\alpha) = u(\beta) = u(\gamma) = 0$ with $0 < a < b < c \le 1$. Then $v(x) = u\left(\frac{m}{m_*}x\right)$ satisfies $v^{(4)} + m^4 v = 0$, $v(0) = v\left(\frac{m_*}{m}\alpha\right) = v\left(\frac{m_*}{m}\beta\right) = v\left(\frac{m_*}{m}\gamma\right) = 0$ and $\frac{m_*}{m}\gamma < \gamma \le 1$.   □

We remark that the interval of nonoscillation can be empty and it is bounded from above by $m = 3\pi\sqrt{2}$. This last property holds as a direct consequence of the previous lemma by using the fact that the function $e^{\frac{mx}{\sqrt{2}}}\sin\left(\frac{mx}{\sqrt{2}}\right)$ is a solution of (1.10.24) that vanishes four times in $[0, 1]$ for $m = 3\pi\sqrt{2}$.

To characterize the values of $m > 0$ for which $[0, 1]$ is a nonoscillation interval, we are interested in finding the infimum of the values m for which exists a solution of (1.10.24) with four zeros in $[0, 1]$. The next lemma allows us to confine our search to the solutions of (1.10.24) that vanish at $x = 1$.

Lemma 1.10.26 ([12, Lemma 3.2]). *Let $m > 0$ be such that there exists a solution $u(x)$ of (1.10.24) such that $u(0) = u(\alpha) = u(\beta) = u(\gamma) = 0$ with $0 < \alpha < \beta < \gamma < 1$. Then m is not the smallest value for which $[0, 1]$ is an interval of oscillation of (1.10.24).*

Proof. Let $v(x) = \frac{u(c\,x)}{c}$. We have $v^{(4)} + (c\,m)^4 v = 0$, $v(0) = v\left(\frac{\alpha}{\gamma}\right) = v\left(\frac{\beta}{\gamma}\right) = v(1) = 0$. Since $\gamma\,m < m$, the result follows.   □

Remark 1.10.27. Notice that in the proof of the previous result we have that $v'(0) = u'(0)$.

Taking $u(1) = 0$ and assuming that $\frac{m}{\sqrt{2}} = n\pi$ for some natural n, we deduce $A = 0$ and, as a consequence, expression (1.10.25) is reduced to

$$u(x) = e^{-n\pi x}\left(Be^{2\pi nx} + D\right)\sin\pi nx.$$

Clearly this function has at most three zeros in $[0, 1]$ when $n = 1$. So, from Lemma 1.10.25, we deduce that interval $[0, 1]$ is nonoscillatory for all $m \in (0, \sqrt{2}\pi]$.

Now, by choosing $D = -2B \neq 0$, we have that for $n = 2$ the previous function vanishes four times in $[0, 1]$. Thus, by using Lemma 1.10.25 again, we know that $[0, 1]$ is oscillatory for all $m \geq 2\sqrt{2}\pi$.

So if $\frac{m}{\sqrt{2}}$ is not a positive multiple of π, we restrict our research to $m \in (\sqrt{2}\pi, 2\sqrt{2}\pi)$. If it is the case, we deduce that in (1.10.25), $u(1) = 0$ if and only if

$$B = -e^{-\sqrt{2}m}\left(A\left(e^{\sqrt{2}m} - 1\right)\cot\left(\frac{m}{\sqrt{2}}\right) + D\right).$$

Now, define $m_0 = \sqrt{2}k_0$, with k_0 defined as in Lemma 1.10.15. It is obvious that $m_0 \approx 5.553$ is the smaller positive solution of the equation

$$\tanh\frac{m}{\sqrt{2}} = \tan\frac{m}{\sqrt{2}}. \tag{1.10.26}$$

Let us consider now the set of solutions u such that $u'(0) = 0$. In this case, the solutions u of (1.10.24) such that $u(0) = u(1) = u'(0) = 0$ follow the expression

$$u(x, m, A) = 2A\sin\left(\frac{mx}{\sqrt{2}}\right)\sinh\left(\frac{mx}{\sqrt{2}}\right)\left(\coth\left(\frac{m}{\sqrt{2}}\right) - \coth\left(\frac{mx}{\sqrt{2}}\right)\right.$$
$$\left. + \cot\left(\frac{mx}{\sqrt{2}}\right) - \cot\left(\frac{m}{\sqrt{2}}\right)\right).$$

It is trivial to see that $\cot\left(\frac{m}{\sqrt{2}}\right) - \coth\left(\frac{m}{\sqrt{2}}\right)$ is a one-to-one map for $m \in (0, m_0]$, so if $0 < m \leq m_0$, we know that

$$\cot\left(\frac{mx}{\sqrt{2}}\right) - \coth\left(\frac{mx}{\sqrt{2}}\right) \neq \cot\left(\frac{m}{\sqrt{2}}\right) - \coth\left(\frac{m}{\sqrt{2}}\right) \quad \text{for all } x \in [0, 1).$$

Since $\frac{mx}{\sqrt{2}} < 2\pi$, the equation $\sin\left(\frac{mx}{\sqrt{2}}\right) = 0$ cannot have more than one solution for $x \in (0, 1)$ and consequently u does not vanish more than three times for $0 < m \leq m_0$.

Now consider the set of solutions u such that $u'(0) \neq 0$. Given u such that $u'(0) \neq 0$, $v(x) = \frac{u(x)}{u'(0)}$ satisfies $v'(0) = 1$ and v has exactly the same zeros as u, so we can just refer to the case $u'(0) = 1$. In consequence, we study the functions given by the expression

$$u(x,m,A) = \frac{2\sinh\left(\frac{mx}{\sqrt{2}}\right)\sin\left(\frac{mx}{\sqrt{2}}\right)}{m}\left[\frac{\sqrt{2}}{2}\left(\coth\left(\frac{mx}{\sqrt{2}}\right) - \coth\left(\frac{m}{\sqrt{2}}\right)\right)\right.$$

$$-Am\left(\left(\cot\left(\frac{mx}{\sqrt{2}}\right) - \cot\left(\frac{m}{\sqrt{2}}\right)\right)\right.$$

$$\left.\left.-\left(\coth\left(\frac{mx}{\sqrt{2}}\right) - \coth\left(\frac{m}{\sqrt{2}}\right)\right)\right)\right].$$ (1.10.27)

Lemma 1.10.28 ([12, Proposition 3.4]). *If $\sqrt{2}\pi < m \le m_0$, then the solutions (1.10.27) of (1.10.24) have at most three zeros.*

Proof. If $u(x,m,A) = 0$ for some $x \in (0,1)$, then it is easy to write A as a function of m and x. Replacing A for $A(x,m)$ in the expression of the first derivative of $u(x,m,A)$ we get that the double zeros of $u(x,m,A)$ belonging to $(0,1)$ must satisfy the condition

$$\frac{\sinh\left(\frac{m(1-x)}{\sqrt{2}}\right)\sinh\left(\frac{mx}{\sqrt{2}}\right)}{\sinh\left(\frac{m}{\sqrt{2}}\right)} = \frac{\sin\left(\frac{m(1-x)}{\sqrt{2}}\right)\sin\left(\frac{mx}{\sqrt{2}}\right)}{\sin\left(\frac{m}{\sqrt{2}}\right)}.$$ (1.10.28)

From (1.10.18) we have that

$$\frac{\sin(m^*x)\sin(m^*(1-s))}{\sin(m^*)} < \frac{\sinh(m^*x)\sinh(m^*(1-s))}{\sinh(m^*)}, \quad \forall x,s \in (0,1), \quad \pi < m^* \le k_0,$$

where k_0 is the smallest positive solution of the equation $\tan k = \tanh k$.

Taking $m^* = \frac{m}{\sqrt{2}}$ and $s = x$, we have that (1.10.28) cannot have a solution if $\sqrt{2}\pi < m \le m_0 \equiv \sqrt{2}k_0$, and, consequently, there are no solutions $u(x,m,A)$ with a double zero in the interval $(0,1)$.

Claim. Given $A < 0$ fixed, there exists $m_A > \sqrt{2}\pi$ close enough to $\sqrt{2}\pi$ such that $u(x,m,A) > 0$ for all $x \in (0,1)$ and all $m \in (\sqrt{2}\pi, m_A)$.

To prove this, we will use the expression (1.10.27):

1. If $x \in \left(0, \frac{\sqrt{2}\pi}{2m}\right]$, we have that $\cot\left(\frac{mx}{\sqrt{2}}\right) - \coth\left(\frac{mx}{\sqrt{2}}\right) < 0$ and $\sin\left(\frac{mx}{\sqrt{2}}\right) > 0$ for all $m \in (\sqrt{2}\pi, 2\sqrt{2}\pi)$. Moreover, there is $\delta_1 > 0$ such that $\cot\left(\frac{m}{\sqrt{2}}\right) - \coth\left(\frac{m}{\sqrt{2}}\right) > 0$ for all $m \in (\sqrt{2}\pi, \sqrt{2}\pi + \delta_1)$. So the function u is positive for some values of $m > \sqrt{2}\pi$ close enough to $\sqrt{2}\pi$.

2. If $x \in \left(\frac{\sqrt{2}\pi}{2m}, \frac{\sqrt{2}\pi}{m} \right)$, we have that $\sin \left(\frac{mx}{\sqrt{2}} \right) > 0$ for all $m \in (\sqrt{2}\pi, 2\sqrt{2}\pi)$ and there exists $\delta_2 > 0$ such that $\cot \left(\frac{mx}{\sqrt{2}} \right) - \cot \left(\frac{m}{\sqrt{2}} \right) \ll 0$ and $\coth \left(\frac{mx}{\sqrt{2}} \right) -$ $\coth \left(\frac{m}{\sqrt{2}} \right) \approx 0$ for all $m \in (\sqrt{2}\pi, \sqrt{2}\pi + \delta_2)$. So again we have $u(x, m, A) > 0$ for m in such an interval.

3. If $x \in \left(\frac{\sqrt{2}\pi}{m}, 1 \right)$, by choosing m close enough to $\sqrt{2}\pi$, we can have the derivative of $- \cot \left(\frac{mx}{\sqrt{2}} \right)$ bounded from below in the given interval by a value as large as necessary. Since $\left| \left(\coth \left(\frac{mx}{\sqrt{2}} \right) \right)' \right| < 1$ and $\sin \left(\frac{mx}{\sqrt{2}} \right) < 0$ it is easy to conclude that $u(x, m, A) > 0$.

Now, since $u(\sqrt{2}\pi/m, m, A) = -2 A \sinh \pi$, the claim is proven.

Let us now focus on the possible double zeros at $x = 1$.

We have

$$u'(1, m, A) = \left(\sqrt{2}Am - 1 \right) \frac{\sin \left(\frac{m}{\sqrt{2}} \right)}{\sinh \left(\frac{m}{\sqrt{2}} \right)} - \sqrt{2}Am \frac{\sinh \left(\frac{m}{\sqrt{2}} \right)}{\sin \left(\frac{m}{\sqrt{2}} \right)},$$

So, for $x = 1$ in order for u to have a double zero, we must have

$$\sqrt{2}Am = \frac{\sin^2 \left(\frac{m}{\sqrt{2}} \right)}{\sin^2 \left(\frac{m}{\sqrt{2}} \right) - \sinh^2 \left(\frac{m}{\sqrt{2}} \right)}. \tag{1.10.29}$$

Since $|\sin(x)| < |\sinh(x)|$ for all $x > 0$, for the previous equality to be true, we must have $A < 0$. In consequence, the function $u(x, m, A)$ has no double zeros in $[0, 1]$ for all $A \geq 0$ and all $m \in (\sqrt{2}\pi, m_0]$. Moreover, since the first derivative at $x = 0$ and $x = 1$ is positive, the number of zeros in $[0, 1]$ is odd. Now, due to the fact that $u(x, m, A)$ has at most three zeros in $[0, 1]$ for all $m \in (0, \sqrt{2}\pi]$, from the continuous dependence with respect to m, we conclude that the function $u(x, m, A)$ has exactly three zeros for all $A \geq 0$ and all $m \in (\sqrt{2}\pi, m_0]$.

For the double zeros at $x = 1$ we can write

$$A(m) = \frac{\sin^2 \left(\frac{m}{\sqrt{2}} \right)}{\sqrt{2}m \left(\sin^2 \left(\frac{m}{\sqrt{2}} \right) - \sinh^2 \left(\frac{m}{\sqrt{2}} \right) \right)}$$

and

$$u''(1, m, A(m)) = - \frac{\sqrt{2}m \left(\cosh \left(\frac{m}{\sqrt{2}} \right) \sin \left(\frac{m}{\sqrt{2}} \right) - \cos \left(\frac{m}{\sqrt{2}} \right) \sinh \left(\frac{m}{\sqrt{2}} \right) \right)}{\sin^2 \left(\frac{m}{\sqrt{2}} \right) - \sinh^2 \left(\frac{m}{\sqrt{2}} \right)},$$

so the double zeros at $x = 1$ must have positive second derivative at $x = 1$ for $m < m_0$.

A careful analysis of the function on the right-hand side of the equality (1.10.29) allows us to conclude that there exist $A_1 < A_0 < 0$ such that:

1. m_0 is a solution of (1.10.29) if and only if $A = A_0$.
2. Equation (1.10.29) has exactly one solution in $(\sqrt{2}\pi, m_0)$ if and only if $A \in [A_0, 0) \cup \{A_1\}$.
3. Equation (1.10.29) has exactly two solutions in $(\sqrt{2}\pi, m_0)$ if and only if $A \in (A_1, A_0)$.
4. Equation (1.10.29) has no solutions in $(\sqrt{2}\pi, m_0)$ if and only if $A < A_1$ or $A \geq 0$.

So, fix $A < 0$. We have three possibilities:

1. If $A < A_1$, we have that the function $u(1, A, m)$ has no double zeros on $[0, 1]$. From the claim and the continuous dependence on the parameter m, we deduce that $u(x, A, m) > 0$ for all $x \in (0, 1)$ and $m \in (\sqrt{2}\pi, m_0]$.
2. Suppose that $A \in [A_0, 0) \cup \{A_1\}$. Let m_1 be the unique value in $(\sqrt{2}\pi, m_0)$ for which $u(1, A, m)$ has a double zero. Since $u''(1, A, m_1) > 0$ the double zero can only provide one extra zero and the solutions $u(x, m, A)$ cannot have more than three zeros in $[0, 1]$ for all $m \in (\sqrt{2}\pi, m_0]$.
3. Suppose that $A \in (A_1, A_0)$. Let $m_1 < m_2$ be the two unique values in $(\sqrt{2}\pi, m_0)$ for which $u(1, A, m)$ has a double zero. As in the previous case, the function $u(x, m, A)$ cannot have more than three zeros in $[0, 1]$ for all $m \in (m_1, m_2)$. Now, the fact that $u''(1, A, m_2) > 0$ implies that $u(x, A, m_2) > 0$ in a neighborhood of $x = 1$. So, if this function takes some negative values in $(0, 1)$, it must have four zeros in $[0, 1]$. The continuous dependence with respect to m implies that the same holds for some values of $m < m_2$, which is not true. As a consequence, $u(x, A, m_2)$ has no zeros in $(0, 1)$ and the double zero at $x = 1$ can only provide one extra zero for $m \in (m_2, m_0]$.

$\square$

In the next result the interval of nonoscillation of operator $T_4[m^4]$ is described.

Theorem 1.10.29 ([12, Theorems 3.5 and 3.6]). $[0, 1]$ *is an interval of nonoscillation for the differential equation* $u^{(4)} + m^4 u = 0$ *if and only if* $m \in (0, m_0]$.

Proof. We have proven in Proposition 1.10.28 that $[0, 1]$ is an interval of nonoscillation for all $m \in (\sqrt{2}\pi, m_0]$. From Lemma 1.10.25 we deduce that the same property holds for all $m \in (0, \sqrt{2}\pi]$.

To see that the interval of nonoscillation is just $(0, m_0]$, it will prove now that for $m > m_0$, there exists a solution $u(x, m, A)$ with a double zero for some $x \in (0, 1)$ and that a small change of the value A provides two zeros in $(0, 1)$ and, consequently, four zeros in $[0, 1]$.

Considering the expression (1.10.28) of the double zeros in $(0, 1)$, let us define

$$f(x, m) = \frac{\sinh\left(\frac{m(1-x)}{\sqrt{2}}\right) \sinh\left(\frac{mx}{\sqrt{2}}\right)}{\sinh\left(\frac{m}{\sqrt{2}}\right)}, \quad g(x, m) = \frac{\sin\left(\frac{m(1-x)}{\sqrt{2}}\right) \sin\left(\frac{mx}{\sqrt{2}}\right)}{\sin\left(\frac{m}{\sqrt{2}}\right)}.$$

If $m_0 < m < 2\sqrt{2}\pi$, we have

$$f\left(\frac{1}{2}, m\right) - g\left(\frac{1}{2}, m\right) = \frac{1}{2}\left(\tanh\left(\frac{m}{2\sqrt{2}}\right) - \tan\left(\frac{m}{2\sqrt{2}}\right)\right) > 0.$$

On the other hand we have

$$f(0, m) - g(0, m) = f'(0, m) - g'(0, m) = 0$$

and

$$f''(0, m) - g''(0, m) = m^2\left(\cot\left(\frac{m}{\sqrt{2}}\right) - \coth\left(\frac{m}{\sqrt{2}}\right)\right) < 0,$$

so there exists x_1 small enough such that $f(x_1, m) - g(x_1, m) < 0$.

Since f and g are continuous functions in the considered domain, for each $m_0 < m < 2\sqrt{2}\pi$ there exists $x_m \in (0, \frac{1}{2})$ such that $f(x_m, m) = g(x_m, m)$ and consequently there is $A_m < 0$ for which the solution $u(x, m, A_m)$ has a double zero in $(0, 1)$.

Let us now see that with the same value of m, a small change of A must provide two zeros in $(0, 1)$. For simplicity let us write the expression (1.10.27) in the compacted form

$$u(x, m, A) \equiv f_1(x, m)(f_2(x, m) - A\, f_3(x, m)).$$

Obviously

$$\frac{\partial u}{\partial A}(x, m, A) = f_1(x, m) f_3(x, m).$$

On the other hand, since $u(x_m, m, A_m) = 0$, we have that

$$f_1(x_m, m) f_2(x_m, m) = -A_m\, f_1(x_m, m)\, f_3(x_m, m).$$

Since $0 < x_m < \frac{1}{2}$ we have that $f_1(x_m, m) > 0$ and $f_2(x_m, m) > 0$. Therefore $f_3(x_m, m) \neq 0$ and we can conclude that $\frac{\partial u}{\partial A}(x_m, m, A_m) \neq 0$. This means that a small change of A_m in one of the directions makes the solution break the $y = 0$ line, providing two zeros for the solution.

Now, from Lemma 1.10.25, we deduce that $[0, 1]$ is an interval of oscillation for all $m > m_0$.   □

So, from the results of Schröder in [49], we know that $T_4[m^4]$ is inverse positive for all $m \in (0, m_0]$.

To see that this property holds only for such values (with $m > 0$), it is necessary to verify that Green's function g_m changes its sign for any $m > m_0$. From Lemma 1.8.33 it is enough to verify this property for $m > m_0$, m close enough to m_0.

This fact has been proved in [12, Theorem 3.10]. To see this, we first note that the expression of Green's function g_m is given in Appendix B. So we can verify that

$$g_m(0,s) = \frac{\partial}{\partial t} g_m(0,s) = 0 \quad \text{for all } s \in [0,1].$$

Moreover

$$\frac{\partial^2}{\partial t^2} g_m(0,s) = -\frac{e^{\frac{m(4-3s)}{\sqrt{2}}}}{\sqrt{2}m\left(\cos\left(\sqrt{2}m\right) + \cosh\left(\sqrt{2}m\right) - 2\right)}$$
$$\left[\left(e^{\sqrt{2}m(s-2)} - e^{2\sqrt{2}m(s-1)}\right)\cos\left(\frac{m(s-2)}{\sqrt{2}}\right)\right.$$
$$+\left(-e^{\sqrt{2}m(s-2)} + e^{2\sqrt{2}m(s-1)}\right)\cos\left(\frac{ms}{\sqrt{2}}\right)$$
$$\left.+\left(e^{\sqrt{2}m(s-2)} - e^{\sqrt{2}m(s-1)} + e^{2\sqrt{2}m(s-1)} - e^{\sqrt{2}m(2s-3)}\right)\sin\left(\frac{ms}{\sqrt{2}}\right)\right].$$

Now, by defining

$$h(s) \equiv \frac{\partial^2}{\partial t^2} g_{m_0/s}(0,s),$$

we deduce that

$$h(1) = h'(1) = h''(1) = 0$$

and

$$h'''(1) = \frac{e^{-\frac{3m_0}{\sqrt{2}}} m_0}{2\left(\cos\left(\sqrt{2}m_0\right) + \cosh\left(\sqrt{2}m\right) - 2\right)^2}$$
$$\left\{3\left(e^{3\sqrt{2}m_0}\left(\sqrt{2} - 2m_0\right) + 4\sqrt{2}e^{\sqrt{2}m_0}\right.\right.$$
$$-4\sqrt{2}e^{2\sqrt{2}m_0} - 2m_0 - \sqrt{2}\right)\cos\left(\frac{m_0}{\sqrt{2}}\right)$$
$$+3e^{\sqrt{2}m_0}\left(2m_0 + e^{\sqrt{2}m_0}\left(2m_0 + \sqrt{2}\right) - \sqrt{2}\right)\cos\left(\frac{3m_0}{\sqrt{2}}\right)$$
$$-\left[e^{3\sqrt{2}m_0}\left(3\sqrt{2} - 2m_0\right) + 2m_0 + e^{\sqrt{2}m_0}\left(38m_0 - 9\sqrt{2}\right) - e^{2\sqrt{2}m_0}\left(38m_0 + 9\sqrt{2}\right)\right.$$
$$\left.\left.+2e^{\sqrt{2}m_0}\left(e^{\sqrt{2}m_0}\left(3\sqrt{2} - 2m_0\right) + 2m_0 + 3\sqrt{2}\right)\cos\left(\sqrt{2}m_0\right) + 3\sqrt{2}\right)\sin\left(\frac{m_0}{\sqrt{2}}\right)\right]\right\}$$
$$\approx 3.4412$$

Thus, we know that there is $\delta > 0$ such that $h(s) < 0$ for all $s \in (1 - \delta, 1)$. In consequence for all $\bar{m} > m_0$ close enough to m_0, there exist $\bar{s} \in (0,1)$ satisfying $\frac{\partial^2}{\partial t^2} g_{\bar{m}}(0,\bar{s}) < 0$ and we conclude that there is $\kappa > 0$ for which

$$g_{\bar{m}}(t,\bar{s}) < 0 \quad \text{for all } t \in (0,\kappa).$$

As a conclusion of the previous results, Theorem 1.8.9 and Lemma 1.4.15, we attain at the following description of the set N_T in the interval $[a,b]$:

Lemma 1.10.30. *Operator $T_4[M]$ is inverse positive in Y_4 if and only if*

$$M \in \left(-\left(\frac{m_1}{b-a} \right)^4, \ 4 \left(\frac{k_0}{b-a} \right)^4 \right]$$

with m_1 given in (1.10.23) and k_0 defined as in Lemma 1.10.15.

As in the simply supported case, it is possible to obtain the optimal values for which operator $T_4[M]$ is inverse positive on bigger sets than Y_4. To this end we present the following formula.

Lemma 1.10.31. *Let $\sigma \in \mathscr{L}^1(J, \mathbb{R})$ and α, β, γ, $\delta \in \mathbb{R}$ be fixed. Assume that operator $T_4[m^4]$ is invertible in Y_4, then the unique solution of problem*

$$T_4[m^4] \, u(t) = \sigma(t), \ a.e. \ \ t \in j, \ \ u(a) = \alpha, \ u(b) = \beta, \ u'(a) = \gamma, \ u'(b) = \delta,$$

is given by the following expression:

$$u(t) = (b-a)^3 \int_a^b g_m \left(\frac{t-a}{b-a}, \frac{s-a}{b-a} \right) \sigma(s) \, ds$$
$$+ \alpha \, w_m(t) + \beta \, w_m(a+b-t) + \gamma \, y_m(t) + \delta \, y_m(a+b-t),$$

where g_m is given in Appendix B, w_m and y_m are defined respectively as the unique solutions of the following problems:

$$T_4[m^4] \, w(t) = 0, \ a.e. \ \ t \in J, \ \ \ w(a) = 1, \ w(b) = w'(a) = w'(b) = 0,$$

and

$$T_4[m^4] \, w(t) = 0, \ a.e. \ \ t \in J, \ \ \ w'(a) = 1, \ w(a) = w(b) = w'(b) = 0.$$

Lemma 1.10.32 ([12, Theorem 3.12]). *Let $M > 0$. Then operator $T_4[M]$ is inverse positive on the set*

$$Z_1 = \{ u \in W^{4,1}(J) : u(a) \geq 0, \ u(b) \geq 0, \ u'(a) = u'(b) = 0 \}$$

if and only if

$$M \in \left(0, \ 4 \left(\frac{\pi}{b-a} \right)^4 \right].$$

Proof. Consider $J = [0, 1]$. Note that $\sqrt{2}\pi < m_0$, so we only need to prove that w_m is positive for $x \in (0, 1)$.

The explicit expression of w_m is

$$\frac{1}{\cos\left(\sqrt{2m}\right)+\cosh\left(\sqrt{2m}\right)-2}\left[\cos\left(\frac{mx}{\sqrt{2}}\right)\cosh\left(\frac{m(x-2)}{\sqrt{2}}\right)-\sin\left(\frac{mx}{\sqrt{2}}\right)\sinh\left(\frac{m(x-2)}{\sqrt{2}}\right)\right.$$
$$\left.+\left(\cos\left(\frac{m(x-2)}{\sqrt{2}}\right)-2\cos\left(\frac{mx}{\sqrt{2}}\right)\right)\cosh\left(\frac{mx}{\sqrt{2}}\right)+\sin\left(\frac{m(x-2)}{\sqrt{2}}\right)\sinh\left(\frac{mx}{\sqrt{2}}\right)\right]$$

and

$$w_m'(x)=\frac{\sqrt{2}m}{\cos\left(\sqrt{2m}\right)+\cosh\left(\sqrt{2m}\right)-2}\left[\left(\cosh\left(\frac{mx}{\sqrt{2}}\right)-\cosh\left(\frac{m(x-2)}{\sqrt{2}}\right)\right)\sin\left(\frac{mx}{\sqrt{2}}\right)\right.$$
$$\left.+\left(\cos\left(\frac{m(x-2)}{\sqrt{2}}\right)-\cos\left(\frac{mx}{\sqrt{2}}\right)\right)\sinh\left(\frac{mx}{\sqrt{2}}\right)\right].$$

It is easy to see that $w_m'(x)<0$ for $m<\sqrt{2}\pi$ which proves that w_m is positive. By differentiating again, we deduce that

$$w_m''(1)=\frac{4m^2\sin\left(\frac{m}{\sqrt{2}}\right)\sinh\left(\frac{m}{\sqrt{2}}\right)}{\cos\left(\sqrt{2m}\right)+\cosh\left(\sqrt{2m}\right)-2},$$

so for $\sqrt{2}\pi<m<2\sqrt{2}\pi$, we have $w_m(x)<0$ for x close enough to 1 and therefore the result is sharp.   □

If $u(0)=u(1)$, the operator is $T_4[m^4]$ is inverse positive in $(0,m_0]$. This property holds because $z_m(x)=w_m(x)+w_m(1-x)$ is nonnegative for $m\le m_z$, where $m_z\approx 6,689$ is the least positive solution of the equation

$$\tanh\left(\frac{m}{2\sqrt{2}}\right)=-\tan\left(\frac{m}{2\sqrt{2}}\right).$$

When the first derivatives are not necessarily zero, the optimal estimation is given in the next result.

Lemma 1.10.33 ([12, Theorem 3.14]). *Let* $M>0$. *Then operator* $T_4[M]$ *is inverse positive on the set*

$$Z_2=\{u\in W^{4,1}(J):u(a)=u(b)=0,\ u'(a)\ge 0\ge u'(b),\}$$

$$M\in\left(0,4\left(\frac{k_0}{b-a}\right)^4\right],$$

with k_0 *defined as in Lemma 1.10.15.*

Proof. In this case, for simplicity, we will not present the long expression for $y_m(x)$, but let us remark that this solution is one of the solutions in (1.10.27) ($J = [0, 1]$) (we have $u(0) = u(1) = 0$ and $u'(0) = 1$). Computing the second derivative, we have that

$$y_m''(1) = \frac{2\sqrt{2}m \left(\cosh\left(\frac{m}{\sqrt{2}}\right) \sin\left(\frac{m}{\sqrt{2}}\right) - \cos\left(\frac{m}{\sqrt{2}}\right) \sinh\left(\frac{m}{\sqrt{2}}\right)\right)}{\cos\left(\sqrt{2}m\right) + \cosh\left(\sqrt{2}m\right) - 2},$$

which is positive for all $m < m_0$ and negative for $m > m_0$ close enough to m_0.

From Proposition 1.10.28, we know that there are no double zeros in $(0, 1)$ for all $m \in (\sqrt{2}\pi, m_0)$. Since for m close enough to 0 we obviously have $y_m(x) > 0$ for all $x \in (0, 1)$. We conclude, by the continuous dependence with respect to m, that y_m is a positive function on $(0, 1)$ for all $m \in (0, m_0]$.

Since g_0 changes sign for $m > m_0$ the result is optimal. $\qquad\square$

One can verify that function y_m takes negative values in $(0, 1)$ near to $x = 1$ for $m > m_0$ close enough. Moreover, in the particular case of $u'(0) = -u'(1) \geq 0$, one can verify that $y_m(t) + y_m(1 - t)$ remains positive for $m \leq 2\sqrt{2}\pi$. However, this fact does not allow to enlarge the interval of positiveness of the operator because m_0 is the maximum value for which Green's function is positive.

As a direct consequence of the two previous lemmas we arrive at the following corollary.

Corollary 1.10.34 ([12, Corollary 3.15]). *Let $M > 0$. Then operator $T_4[M]$ is inverse positive on the set*

$$Z_3 = \{u \in W^{4,1}(J) : u(a) \geq 0,\ u(b) \geq 0,\ u'(a) \geq 0 \geq u'(b)\}$$

if and only if

$$M \in \left(0,\ 4\left(\frac{\pi}{b - a}\right)^4\right].$$

Appendix A
A Green's Function *Mathematica* Package

This appendix is directed to the construction of a *Mathematica* Package valid to calculate the explicit expression of Green's function related to the two-point boundary value problem (1.4.1), where the nth-order linear operator L_n defined on (1.4.3) has constant coefficients. This algorithm has been published in [19] and it can be downloaded from the web page

   http://webspersoais.usc.es/persoais/alberto.cabada/index.html

A.1   The Algorithm

By assuming the uniqueness of solutions of problem (1.4.1) in [19] an algorithm is developed to obtain the expression of Green's function when the operator L_n defined on (1.4.3) has constant coefficients. The method is applied to any two-point boundary conditions by means of the expression of Green's function related to suitable initial problems. By using (1.4.7) for a general initial boundary conditions, the algorithm finds the values of the solution and the successive derivatives up to order $n-1$, for which the two-point boundary conditions hold. Special mention has been made for the periodic case. In this situation the algorithm calculates function (1.4.9).

In the sequel we present the arguments developed in [19, Sect. 2].

First, it is not difficult to verify [42] that Green's function related to the initial value problem

$$L_n y(t) = 0, \ t \in I, \quad y^{(i)}(a) = 0, \ i = 0, \dots, n-1, \tag{A.1.1}$$

A. Cabada, *Green's Functions in the Theory of Ordinary Differential Equations*,
SpringerBriefs in Mathematics, DOI 10.1007/978-1-4614-9506-2,
© Alberto Cabada 2014

is given by

$$\tilde{K}(t,s) = \begin{cases} K(t,s), & \text{if } a \le s \le t, \\ 0, & \text{if } t < s \le b, \end{cases}$$

where

$$K(t,s) := \frac{\begin{vmatrix} y_1(s) & \cdots & y_n(s) \\ y_1'(s) & \cdots & y_n'(s) \\ \vdots & \ddots & \vdots \\ y_1^{(n-2)}(s) & \cdots & y_n^{(n-2)}(s) \\ y_1(t) & \cdots & y_n(t) \end{vmatrix}}{W(y_1,\ldots,y_n)(s)},$$

being $(y_1,\ldots,y_n)$ a fundamental set of solutions of equation $L_n y = 0$ and

$$W(y_1,\ldots,y_n)(s) = \begin{vmatrix} y_1(s) & \cdots & y_n(s) \\ y_1'(s) & \cdots & y_n'(s) \\ \vdots & \ddots & \vdots \\ y_1^{(n-1)}(s) & \cdots & y_n^{(n-1)}(s) \end{vmatrix}$$

its corresponding Wronskian.

To see this, it is enough to check that function $\tilde{K}$ satisfies properties $(g1)$–$(g6)$ in Definition 1.4.1 (See [19, Theorem 2.4] for details).

To obtain the expression of Green's function related to problem (1.4.1), let us consider n continuous functions on J, $c_1,\ldots,c_n$. Now we are going to look for a Green's function of the form

$$g(t,s) = \tilde{K}(t,s) + c_1(s)\,y_1(t) + \cdots + c_n(s)\,y_n(t).$$

It is easy to verify that function g satisfies conditions $(g1)$–$(g5)$ in Definition 1.4.1. Now we must obtain the unique functions $c_1,\ldots,c_n$ for which $(g6)$ is fulfilled, i.e., for each $s \in (a,b)$ we need to verify that

$$U_i(G(\cdot,s)) = 0, \quad \forall\, i = 1,\ldots,n, \quad \forall\, s \in I.$$

By linearity, we have that

$$U_i(G(\cdot,s)) = U_i(\tilde{K}(\cdot,s)) + \sum_{j=1}^{n} c_j(s)\,U_i(y_j), \quad i = 1,\ldots,n,$$

that is, $(c_1(s),\ldots,c_n(s))$ should be a solution of the linear system

$$\begin{pmatrix} U_1(y_1) & \cdots & U_1(y_n) \\ \vdots & \ddots & \vdots \\ U_n(y_1) & \cdots & U_n(y_n) \end{pmatrix} \begin{pmatrix} c_1(s) \\ \vdots \\ c_n(s) \end{pmatrix} = - \begin{pmatrix} U_1(\tilde{K}(\cdot,s)) \\ \vdots \\ U_n(\tilde{K}(\cdot,s)) \end{pmatrix}.$$

Now, since (1.4.1) is uniquely solvable, we have that the system has a unique solution given by

$$\begin{pmatrix} c_1(s) \\ \vdots \\ c_n(s) \end{pmatrix} = - \begin{pmatrix} U_1(y_1) & \cdots & U_1(y_n) \\ \vdots & \ddots & \vdots \\ U_n(y_1) & \cdots & U_n(y_n) \end{pmatrix}^{-1} \begin{pmatrix} U_1(\tilde{K}(\cdot,s)) \\ \vdots \\ U_n(\tilde{K}(\cdot,s)) \end{pmatrix}.$$

Moreover, from this expression, we know that functions $c_1, \ldots, c_n$ are continuous and, therefore, g is Green's function that we are looking for.

From the previous considerations, the problem of deducing the expression of Green's function is reduced to the one of finding a set of $(y_1, \ldots, y_n)$ fundamental solutions of equation $L_n y = 0$. To this end we use the following result for initial value problems, which is proved in [9].

Theorem A.1. *Let r be the unique solution of the initial value problem*

$$u^{(n)}(t) + \sum_{i=0}^{n-1} a_{n-i}\, u^{(i)}(t) = 0, \quad t \in \mathbb{R},$$

$$u^{(i)}(0) = 0, \quad i = 0, \ldots, n-2, \tag{A.1.2}$$

$$u^{(n-1)}(0) = 1.$$

Then, the unique solution of the initial value problem

$$y^{(n)}(t) + \sum_{i=0}^{n-1} a_{n-i}\, y^{(i)}(t) = \sigma(t), \quad t \in J, \tag{A.1.3}$$

$$y^{(i)}(a) = \lambda_i, \quad i = 0, \ldots, n-1,$$

with $\sigma \in \mathcal{L}(J\,\mathbb{R})$ and $\lambda_i \in \mathbb{R}$, $i = 0, \ldots, n-1$, is given by

$$y(t) = \int_a^t r(t-s)\,\sigma(s)\,ds + \sum_{k=0}^{n-1} y_k(t)\,\lambda_k, \tag{A.1.4}$$

where

$$y_k(t) = r^{(n-1-k)}(t-a) + \sum_{j=k+1}^{n-1} a_{n-j}\, r^{(j-k-1)}(t-a), \quad t \in \mathbb{R}, \quad k = 0, \ldots, n-1.$$

$$\tag{A.1.5}$$

We note that in [9] the proof has been done for a continuous function σ. To extend the formula to $\mathcal{L}^1(J, \mathbb{R})$ is immediate.

Now, when we consider the boundary value problem (1.4.1), we will search for a Green's function of the form

$$g(t,s) = \begin{cases} r(t-s) + \sum\limits_{k=0}^{n-1} y_k(t)\,d_k(s), & \text{if} \quad a \le s \le t \le b, \\ \sum\limits_{k=0}^{n-1} y_k(t)\,d_k(s), & \text{if} \quad a \le t < s \le b, \end{cases} \tag{A.1.6}$$

where the continuous real functions d_k are the unknowns.

The expression of d_k came from the verification of the boundary conditions.

$$0 = \sum_{j=0}^{n-1} \left(\alpha_j^i\, y^{(j)}(a) + \beta_j^i\, y^{(j)}(b) \right)$$

$$= \sum_{j=0}^{n-1} \left[\alpha_j^i \int_a^b \sum_{k=0}^{n-1} y_k^{(j)}(a)\,d_k(s)\,\sigma(s)\,ds \right.$$

$$\left. + \beta_j^i \int_a^b r^{(j)}(b-s)\,\sigma(s)\,ds + \beta_j^i \int_a^b \sum_{k=0}^{n-1} y_k^{(j)}(b)\,d_k(s)\,\sigma(s)\,ds \right]$$

$$= \sum_{j=0}^{n-1} \left[\beta_j^i \int_a^b r^{(j)}(b-s)\,\sigma(s)\,ds \right]$$

$$+ \sum_{j=0}^{n-1} \int_a^b d_k(s) \left[\alpha_j^i \sum_{k=0}^{n-1} y_k^{(j)}(a) + \beta_j^i \sum_{k=0}^{n-1} y_k^{(j)}(b) \right] \sigma(s)\,ds$$

$$= \int_a^b \left[\sum_{j=0}^{n-1} \beta_j^i\, r^{(j)}(b-s) + \sum_{k=0}^{n-1} d_k(s)\, U_i(y_k) \right] \sigma(s)\,ds.$$

Since y_k, r and U_i have been previously obtained, by solving the linear system

$$\sum_{k=0}^{n-1} d_k(s)\, U_i(y_k) = -\sum_{j=0}^{n-1} \beta_j^i\, r^{(j)}(b-s), \qquad i = 1,\dots,n, \tag{A.1.7}$$

we obtain the expression of $d_k(s)$ and, therefore, we have the formula for $g(t,s)$.

Notice that system (A.1.7) is equivalent to

$$\begin{pmatrix} U_1(y_0) & \cdots & U_1(y_{n-1}) \\ \vdots & \ddots & \vdots \\ U_n(y_0) & \cdots & U_n(y_{n-1}) \end{pmatrix} \begin{pmatrix} d_0(s) \\ \vdots \\ d_{n-1}(s) \end{pmatrix} = -\begin{pmatrix} \sum\limits_{j=0}^{n-1} \beta_j^1\, r^{(j)}(b-s) \\ \vdots \\ \sum\limits_{j=0}^{n-1} \beta_j^n\, r^{(j)}(b-s) \end{pmatrix}, \tag{A.1.8}$$

which is uniquely solvable.

A.1.1   The **Module** *Environment to Calculate Green's Function*

The purpose of this section, which is essentially [19, Sect. 5], is to construct an algorithm for computing Green's function of problem (1.4.1). Such construction is based on the expression (A.1.6). To arrive at such expression, we must previously find the functions r, y_k, and d_k.

Due to the fact that the function r is the unique solution of the initial value problem (A.1.2), the first step consists on solving such problem.

Once we have this expression, we obtain the expression of the y_k's as the unique solutions of the related problems:

$$y^{(n)}(t) + \sum_{i=0}^{n-1} a_{n-i}\, y^{(i)}(t) = 0, \quad t \in \mathbb{R}, \tag{A.1.9}$$

$$y^{(i)}(a) = 0, \quad i = 0, \ldots, n-1, \, i \neq k, \tag{A.1.10}$$

$$y^{(k)}(a) = 1. \tag{A.1.11}$$

The next step of the algorithm consists of solving the system (A.1.7). In consequence, to ensure the existence and uniqueness of Green's function, we must verify, first, that the matrix of system (A.1.7) is invertible. Otherwise, there is not Green's function and this ends the process.

When the system (A.1.7) is uniquely solvable, once we have obtained its unique solution, d_k, we arrive at the expression of Green's function, $g(t, s)$, by means of the expression (A.1.6) defined in the two triangles $a \leq s < t \leq b$ and $a \leq t < s \leq b$.

As we have noticed in the previous section, the calculations involved in this process are very complicated, so, for higher order equations and several boundary conditions, the resolution may be very slow and the simplifications unavailable. On the other hand, for the particular case of periodic boundary conditions, we only must obtain the function r, defined in (1.4.9), in order to have the expression of Green's function from (1.4.12). The outline of the described algorithm can be seen in the flow diagram of Fig. A.1.

By using the scientific software *Mathematica* 8.0.1.0 and the previously described algorithm, we implemented a program in which, by supplying the order equation, the coefficients of the linear operator and the two-point boundary conditions on the interval J, the related Green's function is calculated.

Using the *Manipulate* environment, and taking advantage of the *Module* that we are going to describe, we will design a "friendly" environment that allows the user to enter data and interact with the program in a simple manner. Moreover, we can see and manipulate both the analytical outputs as well as the graphical results.

In this section we focus our attention on the technical aspects of the *Module* and we leave the aspects of the *Manipulate* to the next subsection.

The *Mathematica Module* that we are going to describe runs once we previously know the following values:

1. The order of the equation (**n**)
2. The coefficients (**vector c**)

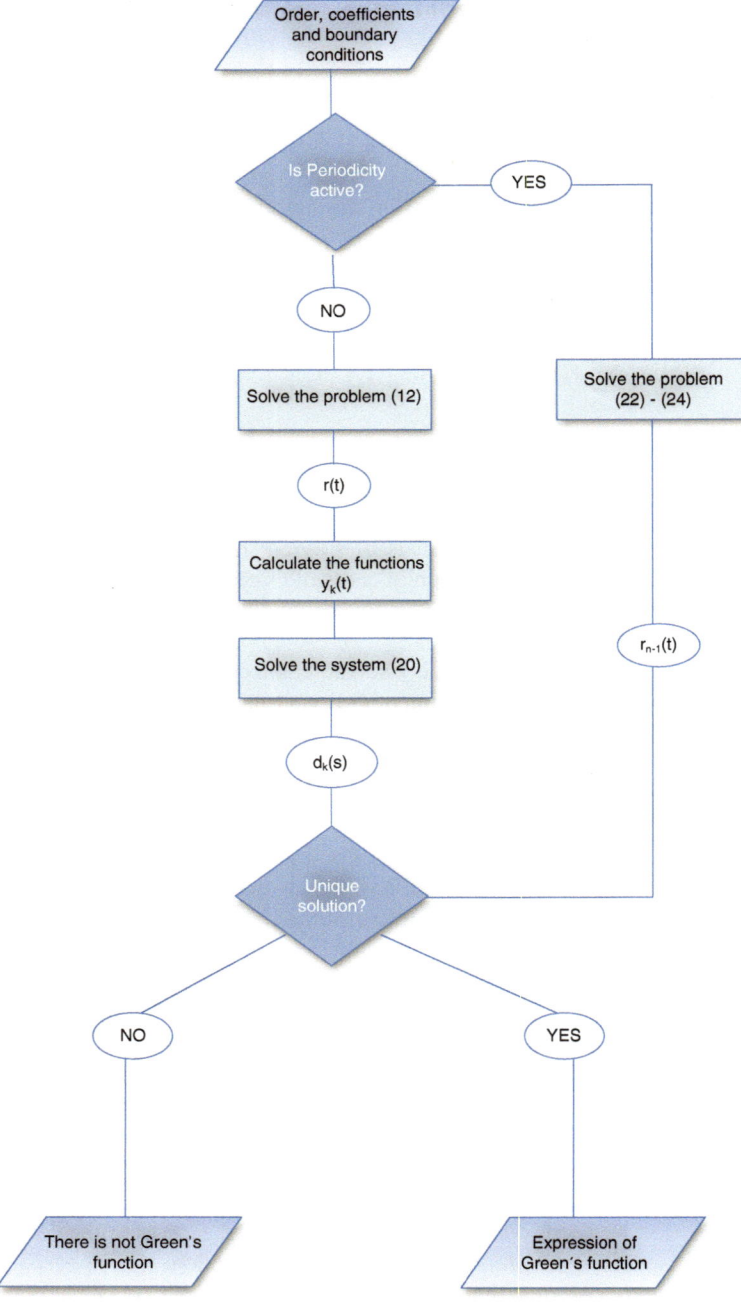

Fig. A.1 Flow diagram of the algorithm

3. The extremes of the interval (**exta** and **extb**)
4. The two-point boundary conditions depending on a and b (**vector cc**)

Moreover we have the option **Periodicity** (False/True).

As a consequence, we can enter the values for the aforementioned variables and simply pick up at the *Module* to run. As we have noticed, in the next section we will explain how to insert this *Module* into the *Manipulate* to enhance the user interaction.

Our *Module* is divided in two algorithms, depending on whether the boundary conditions are, or not, the periodic ones. In this last case, we can choose a more specific algorithm that calculates the function r defined in (1.4.9) Let us start to describe the generic algorithm presented at the beginning of this Appendix, which is the one to be used if we keep the **Periodicity** option as False.

At the beginning, the program checks that both the number of coefficients and of boundary conditions equal the order of the considered equation. Due to the fact that such conditions should be evaluated in the *Module*, in order to find the coefficients of the system described in (A.1.7), these boundary conditions must be a real function and not a vector, as they have been introduced by the keyboard. We have decided to ask for a n-dimensional vector because in this case it is not necessary to enter all the coefficients α_j^i, β_j^i, in functions U_i. It is the *Module* which calculates these coefficients directly from the introduced vector.

To carry out the transformation of each boundary condition into a real function, it is defined as an auxiliary function at the beginning of the *Module* (**aux**). This function saves the vector that contains the boundary conditions as a function that depends on u, ext$a$ and ext$b$. After transforming it into a system, the coefficients that multiply $u^{(j)}(\text{ext}a)$ and $u^{(j)}(\text{ext}b)$ are extracted as α_j^i and β_j^i, respectively. So the function is well defined.

The part of the code that makes this step is showed below. Here **aux** is the auxiliary function that has been previously defined from the vector that contains the boundary conditions:

Do$\left[\text{alfa}[i,j] = \text{Coefficient}\left[\text{aux}[u][\text{ext}a,\text{ext}b][[i]], u^{(j)}[\text{ext}a]\right], \{j, 0, \text{Length}[c] - 1\}, \{i, 1, \text{Length}[c]\}\right]$;

Do$\left[\text{beta}[i,j] = \text{Coefficient}\left[\text{aux}[u][\text{ext}a,\text{ext}b][[i]], u^{(j)}[\text{ext}b]\right], \{j, 0, \text{Length}[c] - 1\}, \{i, 1, \text{Length}[c]\}\right]$;

Once the coefficients α_j^i and β_j^i have been extracted, the functionals are defined U_i, which depends on u, ext$a$, and ext$b$:

$$\text{Do}\left[U_i[u\_][\text{ext}a\_,\text{ext}b\_] = \sum_{j=0}^{\text{Length}[c]-1} \left(\text{alfa}[i,j] * u^{(j)}[\text{ext}a] + \text{beta}[i,j] * u^{(j)}[\text{ext}b]\right), \{i, 1, \text{Length}[c]\}\right].$$

The first step of the algorithm mentioned above is to find the unique solution of the initial value problem (A.1.2). This problem is solved by using the *Mathematica* *DSolve* command. The result is saved as a function r that depends on t.

Denoting by n the order of the equation and by c the vector where the coefficients are saved, (A.1.2) is introduced in the program as

$$\mathtt{ec:=}y^{(\mathtt{Length}[c])}[t] + \sum_{i=1}^{\mathtt{Length}[c]} c[[i]]\, y^{(\mathtt{Length}[c]-i)}[t].$$

The initial value problem (A.1.2) is solved in the sentence

$$\mathtt{DSolve}\Big[\mathtt{Join}\big[\{\mathtt{ec}==0\},\mathtt{Table}\big[y^{(i)}[0]==0,\{i,0,n-2\}\big],\{y^{(n-1)}[0]==1\}\big],y,t\Big];$$

Note that this output is a list, so we need to extract the corresponding part of the function with $y[t]/.\mathrm{DSolve}[\cdots][[1]]$. The input *DSolve* returns the simplified solution, and so on numerous occasions this result shows an expression in which it appears complex numbers when this expression of the solution is the shortest one. We notice that, for higher order equations, it is very usual that *Mathematica* solves the previous equation with dependence on the roots of the characteristic polynomial. In this case the expression of the solution appears as a function of the (unknown for *Mathematica*) corresponding roots. This makes the expression of Green's function impossible to process in practical situations. For this reason the program checks if the word *Root* appears in the expression and, if it is the case, it makes the transformation $c = N[c]$ over the coefficients. This fact implies that *Mathematica* considers such coefficients as a numerical approximation of the corresponding exact numbers, and it makes the next calculations for functions r, y_k and, as consequence, for Green's function G, as numerical approximations too. Our experiments show us that the numerical error is around 10^{-15}.

The second step of the algorithm consists on solving the associated problem (A.1.9)–(A.1.11). The solution is obtained either directly by using *DSolve* or by means of the expression (A.1.5), depending if either all the introduced data is real numbers or there is some parameter. The implementation in *Mathematica* is as follows:

If$[(c \in \text{Reals} \wedge \text{extb} \in \text{Reals} \wedge \text{exta} \in \text{Reals})===\text{True},$
Do $[\mathrm{soluci}[\mathrm{k}] = \mathrm{DSolve}\Big[\mathrm{Join}\big[\{y^{(\mathrm{Length}[c])}[t] + \sum_{i=1}^{\mathrm{Length}[c]} c[[i]]y^{(\mathrm{Length}[c]-i)}[t] == 0\},$
Table $\big[y^{(i)}[0] == 0, \{i, 0, k-2\}\big], \{y^{(k-1)}[0] == 1\},$
Table $\big[y^{(i)}[0] == 0, \{i, k, \mathrm{Length}[c]-1\}\big]\big], y, t\Big];$
$\mathrm{yk}[\mathrm{k}][\mathrm{t\_}] = \mathrm{FullSimplify}[\mathrm{ComplexExpand}[y[t]/.\mathrm{Extract}[\mathrm{soluci}[\mathrm{k}], \{1,1\}]]];\,,$
$\{\mathrm{k}, 1, \mathrm{Length}[c]\}],$
Do $\Big[\mathrm{yk}[\mathrm{k}][\mathrm{t\_}] = r^{(\mathrm{Length}[c]-k)}[t] + \sum_{j=k}^{\mathrm{Length}[c]-1} c[[\mathrm{Length}[c]-j]]r^{(j-k)}[t];\,,$
$\{\mathrm{k}, 1, \mathrm{Length}[c]\}]]$

This different choice follows from our experimental experience with *Mathematica*. We have noticed that when all the variables are real constants, the use of formula (A.1.5) gives us bigger numerical errors than the direct resolution of (A.1.9)– (A.1.11), but (A.1.5) is more adequate if some real parameter is involved in the equations.

The coefficients of the system (A.1.7) are given by the boundary conditions U_i evaluated at y_k. To solve it, we use the *Mathematica* command *Solve*, and the unknown variables are saved on the vector d:

$$\text{Solve}\left[\text{Table}\left[0==\sum_{j=0}^{\text{Length}[c]-1}\text{beta}[i,j]*r^{(j)}[\text{ext}b-s]\right.\right.$$

$$\left.\left.+\sum_{j=1}^{\text{Length}[c]}d_j[s]U_i[y[j]][\text{ext}a,\text{ext}b],\{i,1,\text{Length}[c]\}\right],\text{Table}[d_i[s],\{i,1,\text{Length}[c]\}]\right]$$

We notice that the variables used to solve this system must to be local ones. This is due to the fact that any overlap in the value of the vector d implies that if the dimension of the equation is changed, then the system is incorrectly solved.

Finally, by means of the performed calculations, Green's function is defined as in expression (A.1.6). To make it, we need to extract the coefficients of the solution of the system (A.1.7) returned by *Mathematica*, which is given in a list form. In the following lines of code we present the extraction of such coefficients and the definition of the function h that corresponds to the summation part on both sides of the expression (A.1.6)

$$\text{coef}:=\text{Sort}[\text{Extract}[\text{ecuacion},\{1\}]];$$

$$\text{Do}\left[e[i][s\_]:=d_i[s]/.\text{Extract}[\text{coef},\{i\}];,\{i,1,n\}\right];$$

$$h[t\_,s\_]:=\sum_{i=1}^{\text{Length}[c]}\text{Simplify}\left[e[i][s]\right]y[i][t];$$

The so-called function $h(t,s)$ is the most complicated part of the expression of Green's function and the simplification becomes harder to do.

We recall that this *Module* has two parts: the first one is the generic algorithm, described above, and the second part consists on a specific algorithm to solve equations with periodic boundary conditions. The option **Periodicity** makes the program to run in one way or in another.

In the specific algorithm for periodic boundary conditions it is only needed to solve a boundary value problem and to define Green's function as in (1.4.12).

A.1.2   An Environment Based on **Manipulate**

Once the program has been implemented, the next step is to make a simple environment for the input of the data. To this end, we have programed a *Manipulate* package, which is an interactive environment where users must enter the data through boxes or menus.

While running the program an environment will appear as in Fig. A.2.

In this environment the user must enter the order of the equation, n, that should be a natural number. Then the vector of coefficients must be introduced, which must be written in *Mathematica* format, i.e., between keys and separated by commas. The vector $\{c_1, \ldots, c_n\}$ must have length n and it contains the coefficients accompanying $u^{n-1)}, \ldots, u$.

The coefficients could be real numbers as well as parameters. These parameters can be any lowercase letter that has not been previously used in the calculations (for instance: $j, k, l, m, \ldots$). Notice that the parameter is not considered as a variable by Green's function, so the Graph option does not run in this case.

In the next two boxes the user must insert the endpoints of the interval I. The second one must be strictly bigger than the first one. It is allowed to insert at the endpoints the values a and/or b in a generic way, but in this case the option Graph does not run again, because they are considered as parameters and not as variables.

In the last box the boundary conditions are introduced. They must be inserted as a n-dimensional vector and must take its values at the previously given endpoints a and b. If we choose the option **Periodicity**, the considered boundary conditions are the periodic ones and the program calculates them with the alternative algorithm explained at the beginning of this section. Of course, the periodic boundary conditions can also be introduced in the corresponding box and the program will make the calculations in a generic way.

The **Enter** button controls the execution of the *Manipulate*. While the program is making calculations, it appears pressed.

A.1.3   Final Remarks

Because some of the calculations can exceed the maximum execution time of the machine outputs, we can have an unsatisfactory output that does not return *Manipulate* to the initial situation. In such a case it will be necessary to abort the execution and relaunch the program evaluating again the code.

By making suitable simplifications the output given by the program can be substantially improved. For instance, some commands as **ComplexExpand**, **ExpToTrig**, or **Simplify** may help to get the expression of Green's function without complex numbers. We remark that $G[t, s]$, where Green's function is saved, is a global variable, so once the execution is completed the user can simplify it outside the *Module* by applying the commands that are more suitable in each case.

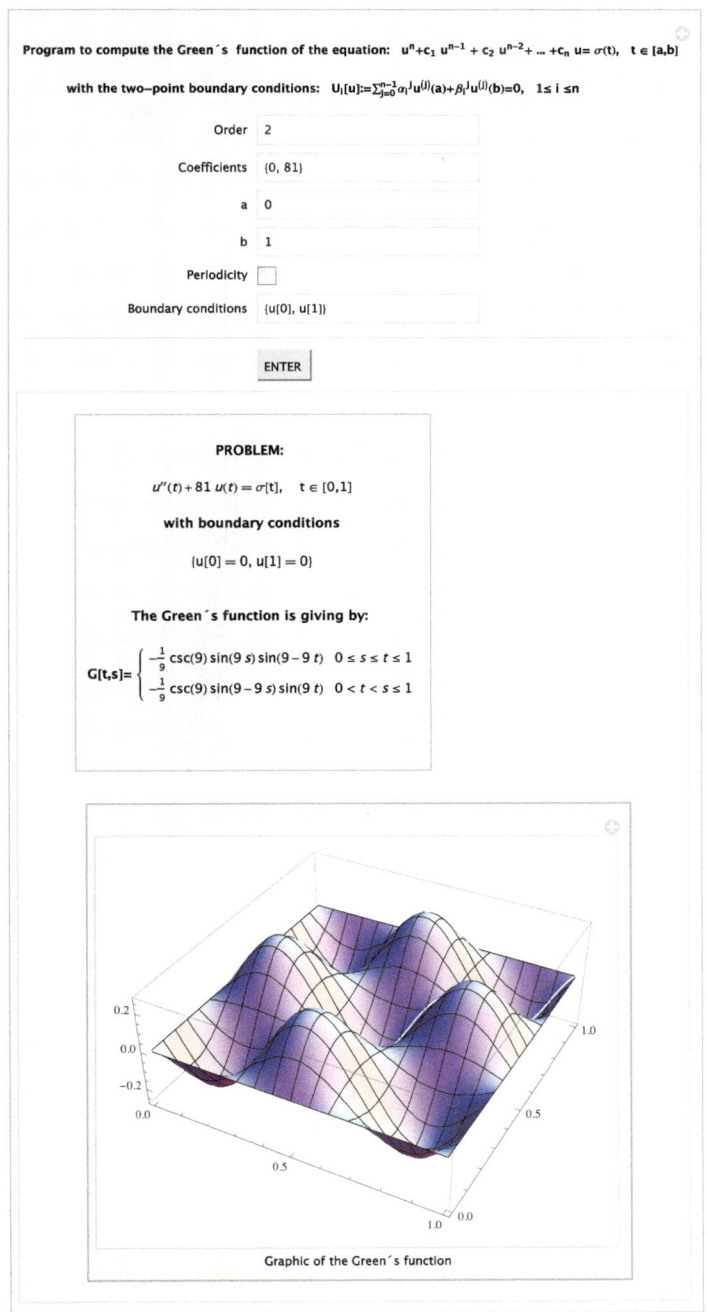

Fig. A.2 Environment for the calculation of Green's function

The complete code of the program is included at the end of the paper as an appendix. Alternatively it can be downloaded from the web page

http://webspersoais.usc.es/persoais/alberto.cabada/index.html

Notice that *Mathematica 8.0.1.0* is recommended in order to run the program. Some trouble could arise if the program is executed in other versions of *Mathematica*.

Appendix B
Expressions of Some Particular Green's Functions

At the Wolfram site a restricted version of the program that has been shown in previous appendix is posted. It may be used without *Mathematica*. In this appendix, the expressions of Green's functions obtained by the *Mathematica* package on that demonstration are given.

B.1 First-Order Problems

Periodic Problem: $u'(t) + m\,u(t) = \sigma(t)$, $\quad t \in [a, b]$, $\quad\quad u(a) = u(b)$.

$$g(t, s, a, b, m) = \frac{1}{e^{bm} - e^{am}} \begin{cases} e^{m(b+s-t)}, & a \le s < t \le b, \\ e^{m(a+s-t)}, & a < t < s \le b. \end{cases}$$

Initial Problem: $u'(t) + m\,u(t) = \sigma(t)$, $\quad t \in [a, b]$, $\quad\quad u(a) = 0$.

$$g(t, s, a, b, m) = \begin{cases} e^{m(s-t)} & a \le s < t \le b, \\ 0 & a < t < s \le b. \end{cases}$$

Terminal Problem: $u'(t) + m\,u(t) = \sigma(t)$, $\quad t \in [a, b]$, $\quad\quad u(b) = 0$.

$$g(t, s, a, b, m) = \begin{cases} 0 & a \le s < t \le b \\ -e^{m(s-t)} & a < t < s \le b. \end{cases}$$

A. Cabada, *Green's Functions in the Theory of Ordinary Differential Equations*,
SpringerBriefs in Mathematics, DOI 10.1007/978-1-4614-9506-2,
© Alberto Cabada 2014

B.2   Second-Order Problems

Periodic Problem: $u''(t) + m^2 u(t) = \sigma(t),\; t \in [a,b],\quad u(a) = u(b),\; u'(a) = u'(b).$

$$g(t,s,a,b,m) = \begin{cases} r(a+t-s,a,b,m), & a \leq s \leq t \leq b \\ r(b+t-s,a,b,m), & a < t \leq s \leq b. \end{cases}$$

with

$$r(t,a,b,m) = -\frac{\csc\left(\frac{1}{2}m(a-b)\right)\cos\left(\frac{1}{2}m(-a+b-2t)\right)}{2m}.$$

Periodic Problem: $u''(t) - m^2 u(t) = \sigma(t),\; t \in [a,b],\quad u(a) = u(b),\; u'(a) = u'(b).$

$$g(t,s,a,b,m) = \begin{cases} r(a+t-s,a,b,m), & a \leq s \leq t \leq b \\ r(b+t-s,a,b,m), & a < t \leq s \leq b. \end{cases}$$

with

$$r(t,a,b,m) = \frac{e^{m(a+t)} + e^{m(b-t)}}{2m\left(e^{am} - e^{bm}\right)}.$$

Dirichlet Problem: $u''(t) + m^2 u(t) = \sigma(t),\quad t \in [a,b],\qquad u(a) = u(b) = 0.$

$$g(t,s,a,b,m) = \begin{cases} h(s,t,a,b,m), & a \leq s \leq t \leq b, \\ h(t,s,a,b,m), & a < t \leq s \leq b. \end{cases}$$

with

$$h(t,s,a,b,m) = -\frac{\sin(m(a-t))\sin(m(b-s))}{m\,\sin(m(a-b))}.$$

Dirichlet Problem: $u''(t) - m^2 u(t) = \sigma(t),\quad t \in [a,b],\qquad u(a) = u(b) = 0.$

$$g(t,s,a,b,m) = \begin{cases} h(s,t,a,b,m), & a \leq s \leq t \leq b, \\ h(t,s,a,b,m), & a < t \leq s \leq b. \end{cases}$$

with

$$h(t, s, a, b, m) = \frac{\left(e^{2mt} - e^{2am}\right)\left(e^{2m(b-s)} - 1\right)e^{m(s-t)}}{2m\left(e^{2am} - e^{2bm}\right)}.$$

Neumann Problem: $u''(t) + m^2 u(t) = \sigma(t), \quad t \in [a, b], \qquad u'(a) = u'(b) = 0.$

$$g(t, s, a, b, m) = \begin{cases} h(s, t, a, b, m), & a \le s \le t \le b, \\ h(t, s, a, b, m), & a < t \le s \le b. \end{cases}$$

with

$$h(t, s, a, b, m) = -\frac{\cos(m(a - t))\cos(m(b - s))}{m \sin(m(a - b))}.$$

Neumann Problem: $u''(t) - m^2 u(t) = \sigma(t), \quad t \in [a, b], \qquad u'(a) = u'(b) = 0.$

$$g(t, s, a, b, m) = \begin{cases} h(s, t, a, b, m), & a \le s \le t \le b, \\ h(t, s, a, b, m), & a < t \le s \le b. \end{cases}$$

with

$$h(t, s, a, b, m) = \frac{\left(e^{2am} + e^{2mt}\right)\left(e^{2m(b-s)} + 1\right)e^{m(s-t)}}{2m\left(e^{2am} - e^{2bm}\right)}.$$

B.3   Third-Order Problems

Periodic Problem: $u'''(t) + m^3 u(t) = \sigma(t), \ t \in [a, b], \quad u^{(i)}(a) = u^{(i)}(b); \ i = 0, 1, 2.$

$$g(t, s, a, b, m) = \begin{cases} r(a + t - s, a, b, m), & a \le s \le t \le b \\ r(b + t - s, a, b, m), & a < t \le s \le b. \end{cases}$$

Where

$$r(t, a, b, m) = -\frac{e^{-m(a+t)}}{3m^2\left(e^{am} - e^{bm}\right)\left(-2e^{\frac{1}{2}m(a+b)}\cos\left(\frac{1}{2}\sqrt{3}m(a - b)\right) + e^{am} + e^{bm}\right)}$$

$$\left[\sqrt{3}e^{\frac{3}{2}m(a+b+t)}\sin\left(\frac{1}{2}\sqrt{3}m(-a + b - t)\right) - \sqrt{3}e^{\frac{1}{2}m(5a+b+3t)}\sin\left(\frac{1}{2}\sqrt{3}m(-a + b - t)\right)\right]$$

$$+\sqrt{3}\sin\left(\frac{1}{2}\sqrt{3}mt\right)e^{\frac{1}{2}m(4a+2b+3t)}+e^{\frac{3}{2}m(a+b+t)}\cos\left(\frac{1}{2}\sqrt{3}m(-a+b-t)\right)$$

$$-e^{\frac{1}{2}m(5a+b+3t)}\cos\left(\frac{1}{2}\sqrt{3}m(-a+b-t)\right)+\left(e^{am}-e^{bm}\right)e^{\frac{1}{2}m(4a+3t)}\cos\left(\frac{1}{2}\sqrt{3}mt\right)$$

$$+e^{m(2a+b)}+e^{m(a+2b)}-2e^{\frac{3}{2}m(a+b)}\cos\left(\frac{1}{2}\sqrt{3}m(a-b)\right)-\sqrt{3}e^{\frac{3}{2}m(2a+t)}\sin\left(\frac{1}{2}\sqrt{3}mt\right)\Bigg].$$

B.4   Fourth-Order Problems

Periodic Problem: $u^{(4)}(t)+m^4\,u(t)=\sigma(t),\ t\in[0,1],\quad u^{(i)}(0)=u^{(i)}(1);\ i=0,1,2,3.$

$$g(t,s,m)=\begin{cases}r(t-s,m), & 0\le s\le t\le 1\\ r(1+t-s,m), & 0\le t\le s\le 1.\end{cases}$$

with

$$r(t,m)=\frac{e^{\frac{mt}{\sqrt{2}}}}{2\sqrt{2}m^3\left(e^{\sqrt{2}m}-2e^{\frac{m}{\sqrt{2}}}\cos\left(\frac{m}{\sqrt{2}}\right)+1\right)}\left[e^{\frac{m}{\sqrt{2}}}\sin\left(\frac{m(1-t)}{\sqrt{2}}\right)\right.$$

$$+e^{\frac{m(1-2t)}{\sqrt{2}}}\sin\left(\frac{m(1-t)}{\sqrt{2}}\right)+e^{\sqrt{2}m(1-t)}\sin\left(\frac{mt}{\sqrt{2}}\right)+\sin\left(\frac{mt}{\sqrt{2}}\right)$$

$$+\left(e^{\frac{m}{\sqrt{2}}}-e^{\frac{m(1-2t)}{\sqrt{2}}}\right)\cos\left(\frac{m(1-t)}{\sqrt{2}}\right)+\left(e^{\sqrt{2}m(1-t)}-1\right)\cos\left(\frac{mt}{\sqrt{2}}\right)\Bigg].$$

Periodic Problem: $u^{(4)}(t)-m^4\,u(t)=\sigma(t),\ t\in[0,1],\quad u^{(i)}(0)=u^{(i)}(1);\ i=0,1,2,3.$

$$g(t,s,m)=\begin{cases}r(t-s,m), & 0\le s\le t\le 1\\ r(1+t-s,m), & 0\le t\le s\le 1.\end{cases}$$

with

$$r(t,m) = \frac{e^{-mt}\left(-e^{2mt} + (e^m - 1)(-e^{mt})\csc\left(\frac{m}{2}\right)\cos\left(\frac{1}{2}m(1 - 2t)\right) - e^m\right)}{4(e^m - 1)m^3}.$$

Simply Supported Conditions:

$$u^{(4)}(t) + m^4 u(t) = \sigma(t),\ t \in [0,1],\ u(0) = u(1) = u''(0) = u''(1) = 0.$$

$$g(t,s,m) = \begin{cases} h(s,t,m), & a \le s \le t \le b, \\ h(t,s,m), & a \le t \le s \le b. \end{cases}$$

with

$$h(t,s,m) = \frac{1}{2\sqrt{2}m^3\left(e^{2\sqrt{2}m} - 2e^{\sqrt{2}m}\cos\left(\sqrt{2}m\right) + 1\right)}e^{-\frac{m(3s+t-6)}{\sqrt{2}}}$$

$$\left(-e^{2\sqrt{2}m(s-1)}\sin\left(\frac{m(s-t-2)}{\sqrt{2}}\right) - e^{\sqrt{2}m(s+t-2)}\sin\left(\frac{m(s-t-2)}{\sqrt{2}}\right)\right.$$

$$+ e^{\sqrt{2}m(2s-3)}\sin\left(\frac{m(s-t)}{\sqrt{2}}\right) + e^{\sqrt{2}m(s+t-1)}\sin\left(\frac{m(s-t)}{\sqrt{2}}\right)$$

$$+ e^{\sqrt{2}m(s-2)}\sin\left(\frac{m(s+t-2)}{\sqrt{2}}\right) + e^{\sqrt{2}m(2s+t-2)}\sin\left(\frac{m(s+t-2)}{\sqrt{2}}\right)$$

$$- e^{\sqrt{2}m(s-1)}\sin\left(\frac{m(s+t)}{\sqrt{2}}\right) - e^{\sqrt{2}m(2s+t-3)}\sin\left(\frac{m(s+t)}{\sqrt{2}}\right)$$

$$+ \left(e^{2\sqrt{2}m(s-1)} - e^{\sqrt{2}m(s+t-2)}\right)\cos\left(\frac{m(s-t-2)}{\sqrt{2}}\right)$$

$$+ \left(e^{\sqrt{2}m(s+t-1)} - e^{\sqrt{2}m(2s-3)}\right)\cos\left(\frac{m(s-t)}{\sqrt{2}}\right) + e^{\sqrt{2}m(s-2)}\cos\left(\frac{m(s+t-2)}{\sqrt{2}}\right)$$

$$- e^{\sqrt{2}m(2s+t-2)}\cos\left(\frac{m(s+t-2)}{\sqrt{2}}\right) - e^{\sqrt{2}m(s-1)}\cos\left(\frac{m(s+t)}{\sqrt{2}}\right)$$

$$\left. + e^{\sqrt{2}m(2s+t-3)}\cos\left(\frac{m(s+t)}{\sqrt{2}}\right)\right).$$

Simply Supported Conditions:

$$u^{(4)}(t) - m^4 u(t) = \sigma(t),\ t \in [0,1],\ u(0) = u(1) = u''(0) = u''(1) = 0.$$

$$g(t,s,m) = \begin{cases} h(s,t,m), & 0 \le s \le t \le 1, \\ h(t,s,m), & 0 \le t \le s \le 1. \end{cases}$$

with

$$h(t,s,m)$$

$$= \frac{1}{4\left(e^{2m}-1\right)m^3}\left(e^{-m(s+t-2)}\left(e^{2m(s+t-1)}\right.\right.$$

$$\left.\left.-2\csc(m)\left(e^{m(s+t-2)}-e^{m(s+t)}\right)\sin(m-ms)\sin(mt)-e^{2m(s-1)}-e^{2mt}+1\right)\right).$$

Clamped Beam Conditions:

$$u^{(4)}(t)+m^4\,u(t)=\sigma(t),\ t\in[0,1],\ u(0)=u(1)=u'(0)=u'(1)=0.$$

$$g(t,s,m)=\begin{cases}h(s,t,m),&0\le s\le t\le 1,\\h(t,s,m),&0\le t\le s\le 1.\end{cases}$$

with

$$h(t,s,m)$$

$$=\frac{e^{-\frac{m(3s+t-6)}{\sqrt{2}}}}{2\sqrt{2}m^3\left(-4e^{\sqrt{2}m}+e^{2\sqrt{2}m}+2e^{\sqrt{2}m}\cos\left(\sqrt{2}m\right)+1\right)}$$

$$\left[\left(-e^{\sqrt{2}m(s-2)}\sin\left(\frac{m(s-2)}{\sqrt{2}}\right)+e^{2\sqrt{2}m(s-1)}\sin\left(\frac{m(s-2)}{\sqrt{2}}\right)\right.\right.$$

$$+e^{\sqrt{2}m(s-1)}\sin\left(\frac{ms}{\sqrt{2}}\right)-e^{\sqrt{2}m(2s-3)}\sin\left(\frac{ms}{\sqrt{2}}\right)$$

$$+\left(e^{\sqrt{2}m(s-2)}+e^{2\sqrt{2}m(s-1)}\right)\cos\left(\frac{m(s-2)}{\sqrt{2}}\right)$$

$$\left.+\left(-2e^{\sqrt{2}m(s-2)}+e^{\sqrt{2}m(s-1)}-2e^{2\sqrt{2}m(s-1)}+e^{\sqrt{2}m(2s-3)}\right)\cos\left(\frac{ms}{\sqrt{2}}\right)\right)$$

$$\left(\left(e^{\sqrt{2}mt}-1\right)\cos\left(\frac{mt}{\sqrt{2}}\right)-\left(e^{\sqrt{2}mt}+1\right)\sin\left(\frac{mt}{\sqrt{2}}\right)\right)$$

$$-2\left(e^{\sqrt{2}mt}-1\right)\sin\left(\frac{mt}{\sqrt{2}}\right)$$

$$\left(\left(e^{\sqrt{2}m(s-2)}-e^{\sqrt{2}m(s-1)}+e^{2\sqrt{2}m(s-1)}-e^{\sqrt{2}m(2s-3)}\right)\sin\left(\frac{ms}{\sqrt{2}}\right)\right.$$

$$\left.\left.+\left(e^{\sqrt{2}m(s-2)}-e^{2\sqrt{2}m(s-1)}\right)\cos\left(\frac{m(s-2)}{\sqrt{2}}\right)+\left(e^{2\sqrt{2}m(s-1)}-e^{\sqrt{2}m(s-2)}\right)\cos\left(\frac{ms}{\sqrt{2}}\right)\right)\right].$$

Clamped Beam Conditions:

$$u^{(4)}(t) - m^4 u(t) = \sigma(t), \; t \in [0,1], \; u(0) = u(1) = u'(0) = u'(1) = 0.$$

$$g(t,s,m) = \begin{cases} h(s,t,m), & 0 \le s \le t \le 1, \\ h(t,s,m), & 0 \le t \le s \le 1. \end{cases}$$

with

$h(t,s,m)$

$$= \frac{e^{m(s-1)}}{8m^3 \left((e^{2m}+1)\cos(m) - 2e^m \right)} \left[\left(e^{-mt} + e^{mt} - 2\cos(mt) \right) \right.$$

$$\left(e^{-2m(s-1)} + e^{m(3-2s)} \right) \sin(m) - 2e^{-m(s-2)} \sin(ms) - e^{-m(s-3)} \sin(m - ms)$$

$$- e^{m-ms} \sin(m - ms) + \left(e^m - e^{m(3-2s)} \right) \cos(m)$$

$$+ \left(e^{2m} - 1 \right) e^{m-ms} \cos(m - ms) - e^{2m} + e^m \sin(m))$$

$$- \left(e^{-mt} - e^{mt} + 2\sin(mt) \right) \left(e^{-2m(s-1)} - e^{m(3-2s)} \sin(m) + e^{-m(s-3)} \sin(m-ms) \right.$$

$$- e^{m-ms} \sin(m - ms) - \left(e^{m(3-2s)} + e^m \right) \cos(m) + 2e^{-m(s-2)} \cos(ms)$$

$$\left. \left. - e^{-m(s-3)} \cos(m - ms) - e^{m-ms} \cos(m - ms) + e^{2m} + e^m \sin(m)) \right].$$

Glossary

$J = [a, b] = \{x \in \mathbb{R},\ a \le x \le b\}$. In some particular situations J corresponds to the intervals $[0, 1]$, $[0, \pi]$ or $[0, R]$.

I_n: Identity matrix of dimension n.

$\|f\|_\infty = \sup_{t \in J} \{|f(t)|\}$.

$\|f\|_p = \left(\int_a^b |f(t)|^p \, dt \right)^{1/p}$.

$\mathscr{M}_{n \times n}$: Set of square matrices of dimension equals to n.

$\ker(L)$: Kernel of operator L, i.e., the set of x such that $L\,x = 0$.

$\det(M)$: Determinant of the matrix M.

$\text{rank}\ (M)$: Rank of the matrix M.

M^{-1}: Inverse matrix of the matrix M.

M^T: Transposed matrix of the matrix M.

T^*: Adjoint of operator T.

$\mathscr{L}^p(J, \mathbb{R}) = \{f : J \to \mathbb{R},\ f \text{ is measurable on } J \text{ and } \int_a^b |f(t)|^p \, dt < \infty\}$.

$\mathscr{C}^m(J, \mathbb{R}) = \{f : J \to \mathbb{R},\ f^{(j)} \text{ is continuous on } J \text{ for all } j \in \{0, \ldots, m\}\ \}$.

$\mathscr{AC}(J, \mathbb{R}) = \{f : J \to \mathbb{R},\ f \in \mathscr{C}(J, \mathbb{R}),\ f' \in \mathscr{L}^1(J, \mathbb{R}) \text{ and } f(t) - f(s) = \int_s^t f'(r) \, dr \text{ for all } t, s \in J\}$.

$\mathscr{W}^{m,p}(J, \mathbb{R}) = \{f : J \to \mathbb{R},\ f \in \mathscr{C}^{m-1}(J, \mathbb{R}),\ f^{(m-1)} \in \mathscr{AC}(J, \mathbb{R}) \text{ and } f^{(m)} \in \mathscr{L}^p(J, \mathbb{R})\}$.

$\mathscr{L}^p(J, \mathbb{R}^n) = \{f \equiv (f_1, \ldots, f_n) : J \to \mathbb{R}^n,\ f_j \in \mathscr{L}^p(J, \mathbb{R}) \text{ for all } j \in \{0, \ldots, n\}\ \}$.

$\mathscr{C}^m(J, \mathbb{R}^n) = \{f \equiv (f_1, \ldots, f_n) : J \to \mathbb{R}^n,\ f_j \in \mathscr{C}^m(J, \mathbb{R}) \text{ for all } j \in \{0, \ldots, n\}\}$.

$\mathscr{AC}(J, \mathbb{R}^n) = \{f \equiv (f_1, \ldots, f_n) : J \to \mathbb{R}^n,\ f_j \in \mathscr{AC}(J, \mathbb{R}) \text{ for all } j \in \{0, \ldots, n\}\}$.

A. Cabada, *Green's Functions in the Theory of Ordinary Differential Equations*,
SpringerBriefs in Mathematics, DOI 10.1007/978-1-4614-9506-2,
© Alberto Cabada 2014

$$\mathscr{W}^{m,p}(J,\mathbb{R}^n) = \{f \equiv (f_1,\ldots,f_n) : J \to \mathbb{R}^n, \; f_j \in \mathscr{W}^{m,p}(J,\mathbb{R})$$
$$\text{for all } j \in \{0,\ldots,n\}\}.$$

$$\mathscr{L}^p(J,\mathscr{M}_{n\times n}) = \{f \equiv (f_{i,j})_{i,\,j\in\{1,\ldots,n\}} : J \to \mathscr{M}_{n\times n}, \; f_{i,j} \in \mathscr{L}^p(J,\mathbb{R})$$
$$\text{for all } i,\; j \in \{0,\ldots,n\}\}.$$

References

1. Afuwape, A.U., Omari, P., Zanolin, F.: Nonlinear perturbations of differential operators with nontrivial kernel and applications to third-order periodic boundary value problems. J. Math. Anal. Appl. **143**, 35–56 (1989)
2. Appell, J., Zabrejko, P.P.: Nonlinear superposition operators. Cambridge Tracts in Mathematics, vol. 95. Cambridge University Press, Cambridge (1990)
3. Bernfeld, S.R., Lakshmikantham, V.: An introduction to nonlinear boundary value problems. Mathematics in Science and Engineering, vol. 109. Academic Press, New York (1974)
4. Cabada, A.: The method of lower and upper solutions for second, third, fourth, and higher order boundary value problems. J. Math. Anal. Appl. **185**, 302–320 (1994)
5. Cabada, A.: The monotone method for first-order problems with linear and nonlinear boundary conditions. Appl. Math. Comput. **63**, 163–186 (1994)
6. Cabada, A.: The method of lower and upper solutions for nth-order periodic boundary value problems. J. Appl. Math. Stoch. Anal. **7**, 33–47 (1994)
7. Cabada, A.: The method of lower and upper solutions for third-order periodic boundary value problems. J. Math. Anal. Appl. **195**, 568–589 (1995)
8. Cabada, A.: The monotone method for third order boundary value problems. In: Proceedings of the World Congress of Nonlinear Analysts, vol. I, pp. 211–221, Aug 1992. Walter de Gruyter, Tampa (1996)
9. Cabada, A.: Maximum principles for third-order initial and terminal value problems. In: Differential & Difference Equations and Applications, pp. 247–255. Hindawi Publishing Corporation, New York (2006)
10. Cabada, A.: An overview of the lower and upper solutions method with nonlinear boundary value conditions. Bound. Value Prob. **2011**, 18 (2011) Article ID 893753
11. Cabada, A., Cid, J.A.: On the sign of the Green's function associated to Hill's equation with an indefinite potential. Appl. Math. Comput. **205**, 303–308 (2008)
12. Cabada, A., Enguiça, R.: Positive solutions of fourth order problems with clamped beam boundary conditions. Nonlinear Anal. **74**, 3112–3122 (2011)
13. Cabada, A., Lois, S.: Maximum principles for fourth and sixth order periodic boundary value problems. Nonlinear Anal. **29**(10), 1161–1171 (1997)
14. Cabada, A., Lois, S.: Existence results for nonlinear problems with separated boundary conditions. Nonlinear Anal. **35**, 449–456 (1999)
15. Cabada, A., Nieto, J.J.: Approximation of solutions for second order boundary value problems. Bull. Classe Sci. Acad. Roy. Bel. 6$^{\underline{e}}$ Sér. II **10–11**, 287–311 (1991)
16. Cabada, A., Sanchez, L.: A positive operator approach to the Neumann problem for a second order ordinary differential equation. J. Math. Anal. Appl. **204**(3), 774–785 (1996)

17. Cabada, A., Tojo, F.A.: Comparison results for first order linear operators with reflection and periodic boundary value conditions. Nonlinear Anal. **78**, 32–46 (2013)
18. Cabada, A., Cid, J.A., Sanchez, L.: Positivity and lower and upper solutions for fourth order boundary value problems. Nonlinear Anal. **67**, 1599–1612 (2007)
19. Cabada, A., Cid, J.A., Máquez-Villamarín, B.: Computation of Green's functions for boundary value problems with Mathematica. Appl. Math. Comput. **219**(4), 1919–1936 (2012)
20. Coddington, E.A., Levinson, N.: Theory of Ordinary Differential Equations. McGraw-Hill, New Delhi (1987)
21. Coppel, W.A.: Disconjugacy. In: Lecture Notes in Mathematics, vol. 220. Springer, Berlin (1971)
22. De Coster, C., Habets, P.: An overview of the method of lower and upper solutions for ODEs. Nonlinear analysis and its applications to differential equations (Lisbon, 1998). Progress in Nonlinear Differential Equations and their Applications, vol. 43, pp. 3–22. Birkhäuser, Boston (2001)
23. De Coster, C., Habets, P.: The lower and upper solutions method for boundary value problems. Handbook of Differential Equations, pp. 69–160, Elsevier/North-Holland, Amsterdam (2004)
24. De Coster, C., Habets, P.: Two-point boundary value problems: lower and upper solutions. Mathematics in Science and Engineering, vol. 205. Elsevier B.V., Amsterdam (2006)
25. Duffy, D.G.: Green's functions with applications. Studies in Advanced Mathematics. Chapman & Hall/CRC, Boca Raton (2001)
26. Fabry, C., Habets, P.: Upper and lower solutions for second-order boundary value problems with nonlinear boundary conditions. Nonlinear Anal. **10**(10), 985–1007 (1986)
27. Fried, H.M.: Green's Functions and Ordered Exponentials. Cambridge University Press, Cambridge (2002)
28. Hartman, P.: Ordinary Differential Equations. Wiley, New York (1964)
29. Karlin, S.: Positive operators. J. Math. Mech. **8**(6), 907–937 (1959)
30. Karlin, S.: The existence of eigenvalues for integral operators. Trans. Am. Math. Soc. **113**, 1–17 (1964)
31. Kythe, P.: Green's functions and linear differential equations. Theory, applications, and computation. Chapman & Hall/CRC Applied Mathematics and Nonlinear Science Series. CRC Press, Boca Raton (2011)
32. Ladde, G.S., Lakshmikantham, V., Vatsala, A.S.: Monotone Iterative Techniques for Nonlinear Differential Equations. Pitman, Boston (1985)
33. Lloyd, N.G.: Degree theory. Cambridge Tracts in Mathematics, vol. 73. Cambridge University Press, Cambridge (1978)
34. Mawhin, J.: Points fixes, points critiques et problèmes aux limites. (French) Séminaire de Mathématiques Supérieures, vol. 92, p. 162. Presses de l'Université de Montréal, Montreal (1985)
35. Mawhin, J.: Twenty years of ordinary differential equations through twelve Oberwolfach meetings. Results Math. **21**(1–2), 165–189 (1992)
36. Mawhin, J.: Boundary value problems for nonlinear ordinary differential equations: from successive approximations to topology. Development of Mathematics 1900–1950 (Luxembourg, 1992), pp. 443–477. Birkhäuser, Basel (1994)
37. Mawhin, J.: Bounded solutions of nonlinear ordinary differential equations. Non-linear analysis and boundary value problems for ordinary differential equations (Udine). CISM Courses and Lectures, vol. 371, pp. 121–147. Springer, Vienna (1996)
38. Melnikov, Y.A.: Green's functions and infinite products. Bridging the Divide. Birkhäuser/Springer, New York (2011)
39. Melnikov, Y.A., Melnikov, M.Y.: Green's functions. Construction and applications. de Gruyter Studies in Mathematics, vol. 42. Walter de Gruyter & Co., Berlin (2012)
40. Müller, M.: Über das Fundamentaltheorem in der theorie der gewöhnlichen differentialgleichungen. Math. Z. **26**, 619–649 (1926)

41. Nkashama, M.N.: A generalizaded upper and lower solutions method and multiplicity results for nonlinear first-order ordinary differential equations. J. Math. Anal. Appl. **140**, 381–395 (1989)

42. Novo, S., Obaya, R., Rojo, J.: Equations and Differential Systems (in Spanish). McGraw-Hill, New York (1995)

43. Omari, P., Trombetta, M.: Remarks on the lower and upper solutions method for second and third-order periodic boundary value problems. Appl. Math. Comput. **50**, 1–21 (1992)

44. Perron, O.: Ein neuer existenzbeweis für die integrale der differentialgleinchung $y' = f(t, y)$. Math. Ann. **76**, 471–484 (1915)

45. Picard, E.: Mémoire sur la théorie des équations aux derivés partielles et las méthode des approximations succesives. J. Math. **6**, 145–210 (1890)

46. Picard, E.: Sur l'application des métodes d'approximations succesives à l'étude de certains équations différentielles ordinaires. J. Math. **9**, 217–271 (1893)

47. Renardy, M., Rogers, R.C.: An introduction to partial differential equations. Texts in Applied Mathematics, vol. 13, 2nd edn. Springer, New York (2004)

48. Rudin, W.: Principles of Mathematical Analysis. McGraw-Hill, New York (1976)

49. Schröder, J.: On linear differential inequalities. J. Math. Anal. Appl. **22**, 188–216 (1968)

50. Schröder, J.: Operator inequalities. Mathematics in Science and Engineering, vol. 147. Academic [Harcourt Brace Jovanovich, Publishers], New York (1980)

51. Scorza Dragoni, S.: Il problema dei valori ai limiti estudiato in grande per le equazione differenziale del secondo ordine. Math. Ann. **105**, 133–143 (1931)

52. Šeda, V., Nieto, J.J., Gera, M.: Periodic boundary value problems for nonlinear higher order ordinary differential equations. Appl. Math. Comput. **48**, 71–82 (1992)

53. Şeremet, V.D.: Handbook of Green's Functions and Matrices. With 1 CD-ROM (Windows and Macintosh). WIT Press, Southampton (2003)

54. Stakgold, I., Holst, M.: Green's functions and boundary value problems, 3rd edn. Pure and Applied Mathematics (Hoboken). Wiley, Hoboken (2011)

55. Zeidler, E.: Nonlinear Functional Analysis and its Applications. I. Fixed-Point Theorems. Translated from the German by Peter R. Wadsack. Springer, New York (1986)

Index

A. Cabada, *Green's Functions in the Theory of Ordinary Differential Equations*,
SpringerBriefs in Mathematics, DOI 10.1007/978-1-4614-9506-2,
© Alberto Cabada 2014